Lecture Notes in Mathematics

A collection of informal reports and seminars
Edited by A. Dold, Heidelberg and B. Eckmann, Zürich

223

Ulrich Felgner

Universität Heidelberg, Heidelberg/Deutschland

Models of ZF-Set Theory

Springer-Verlag
Berlin · Heidelberg · New York 1971

AMS Subject Classifications (1970): 02 K 05, 02 K 15, 02 K 20, 04-02, 04 A 25

ISBN 3-540-05591-6 Springer Verlag Berlin · Heidelberg · New York
ISBN 0-387-05591-6 Springer Verlag New York · Heidelberg · Berlin

Offsetdruck: Julius Beltz, Hemsbach/Bergstr.

PREFACE

This set of notes is part and parcel of a series of lectures given in 1970 from February up to June at the Department of Mathematics at the State University at Utrecht and they were intended only as an aid to the people attending those lectures. They were written as the course developed. In fact, in spite of their official looking aspect, they are no more than prelecture scribblings. In most cases complete and detailed proofs are given in these notes but in some few places there are only short indications to the proof. This occurred when a result was only slighty touched in order to round up the presentation.

It was the aim of these lectures to provide an exposition of some of the basic techniques and results of the theory of models of Set Theory. This theory covers a wide field and it is not possible to compress it into one series of lectures. We have chosen as our theme the construction of Gödel's model L of constructible sets, the construction of Fraenkel-Mostowski-Specker models (containing ungrounded sets like $x = \{x\}$) and Cohen-generic models. As an introduction (chapter I), the axiomatization of Zermelo-Fraenkel is given, some basic concepts and Lévy's principles of reflection. Chapter II contains Gödel's relative consistency proof for the axiom of choice, the generalized continuum hypothesis and the axiom of constructibility. Chapters III and IV contain the methods of Fraenkel-Mostowski-Specker and of P.Cohen in a general setting and various applications. Many of these applications have not yet appeared in print. As an aid to the reader several informal discussions and explanations are included. Although we have attempted to reduce mistakes and obscurities to a minimum, we should be glad to have our attention drawn to any indiscretion the reader may discover in the text.

To finish this preface I wish to express my gratitude to D.van Dalen and the Department of Mathematics at Utrecht for the kind invitation to spend a year at this Institute. Thanks are due to R.de Vrijer and K.Rasmussen for correcting several misprints and to Helène Keller for typing the manuscript.

Heidelberg, July 5, 1971

CONTENTS

CHAPTER I

Prerequisites

In this short chapter we list those topics, mostly taken from the lower predicate calculus and Peano Arithmetic, that we expect our readers to be already familiar with. For more details the reader is referred to

[15] S. FEFERMAN: Arithmetization of metamathematics in a general setting; Fund. Math. 49 (1960) p. 35-92.

[60] E. MENDELSON: Introduction to Mathematical Logic; v. Nostrand Company, Princeton-New York-London 1966 (3rd printing).

[75] J.R. SHOENFIELD: Mathematical Logic; Addison-Wesley Publ.Comp. 1967.

All our considerations about formal languages, their syntax and semantics are carried out in a certain underlying Metatheory. Our metalanguage will be english enriched by some mathematical and logical symbols such as:

$$\sim, \&, \dot{\vee}, \Rightarrow, \Leftrightarrow, \forall, \exists, \triangleq$$

(negation, conjunction, disjunction, implication, equivalence, universal quantification, existential quantification and equality).

It is understood that the use of these symbols is known, a semantics (id est: set theoretical interpretation) is not given. The symbol $\in$ denotes "membership". We adopt naive set theory as our metatheory (id est: Zermelo-Fraenkel set theory as formulated in the above meta-language). The universe of sets as given by our naive set theory will be called sometimes "the real world" and objects, operations and relations in it shall get sometimes the adjective "actual" in order to distinguish them from the corresponding objects, operations and relations of some "object-theory". Hence $\in$ is the actual membership relation and $\omega \triangleq \omega_0$ is the actual set of (actual) natural numbers.

A) RECURSIVE FUNCTIONS

We adopt the usual definitions of primitive recursive and recursive (id est: general recursive) functions of ω into ω (see e.g. Mendelson [60] p. 120-121). We recall briefly the definitions:

A function f is said to be primitive recursive iff it can be obtained from the functions $Z(x) = 0$, $N(x) = x+1$, $U_i^n(x_1,\ldots,x_n) = x_i$, the initial functions, by any finite number of substitutions and recursion. If f can be obtained in this way but with some finite application of the μ-Operator, then f is called recursive.

A relation $R(x_1,\ldots,x_n)$ between natural numbers is said to be primitive recursive (recursive) iff its characteristic function

$$\chi_R(x_1,\ldots,x_n) = \begin{cases} 0 \text{ if } \sim R(x_1,\ldots,x_n) \\ 1 \text{ if } R(x_1,\ldots,x_n) \end{cases}$$

is primitive recursive (recursive, respectively).

A subset A of ω is called a recursively enumerable (r.e.) set iff B is either empty or the range of a recursive function. A subset B of ω is recursive iff both B and its complement ω-B are r.e. It is well known that A is r.e. iff "$x \in A$" is expressible in the form $(\exists y)(R(x,y))$ for some recursive R (one can allow R here to be primitive recursive).

A function f is primitive recursive (recursive) in the relation R iff f can be obtained from the initial functions together with the characteristic function χ_R of R by any finite number of substitutions and recursion (and application of the μ-Operator, respectively).

B) FORMAL THEORIES

A formal theory $\underline{T}$ is a triple $\langle \underline{L},\underline{C},\underline{V} \rangle$ where $\underline{L}$ is a formal language, $\underline{C}$ is a set of consequence operations and $\underline{V}$ a set of sentences of $\underline{L}$ closed under $\underline{C}$, called the valid sentences of $\underline{T}$. Here we shall discuss only those theories $\underline{T}$ where $\underline{C}$ is the first-order predicate calculus and $\underline{L}$ is an elementary language. Such languages $\underline{L}$ can be defined as abstract algebras whose elements are just the well formed formulas, see e.g.

[70] H. RASIOWA-R.SIKORSKI: The Mathematics of Metamathematics; Monografie Matematyczne, vol. 41, Warszawa 1963.

Let K an alphabet consisting of one sort of variables $v_0, v_1, \ldots, v_n, \ldots (n \in \omega)$, countably many constants c_n, countably many primitive predicates π_n and logical symbols $\neg, \wedge, \bigwedge$ (negation conjunction, universal quantification) the other logical symbols: $\vee, \rightarrow, \leftrightarrow, \bigvee$ (disjunction, implication, equivalence, existential quantification) are definable in terms of $\neg, \wedge, \bigwedge$). Remark that "countable" means "finite or countably infinite". From these signs of K the set At_K of atomic formulae, the set T_K of terms of K and the set $\underline{L}_K$ of all (well formed) formulae are obtained as usual. Certain formulae of $\underline{L}_K$ are called logical axioms (in K) and in order to be definite we take those defined in Mendelson [60], p.57. The set C of rules of inference consists of the "modus ponens" and the "Generalization" (cf. [60] p. 57).

For any ordered pair $\langle \underline{L},\underline{A} \rangle$, where $\underline{L}$ is a first-order language as just described, and $\underline{A}$ is a set of sentences from $\underline{L}$, one can define the proof relation $PRF_{\langle \underline{L},\underline{A} \rangle}$:

$PRF_{\langle \underline{L},\underline{A} \rangle}(\Phi,S)$ holds iff S is a finite sequence of formulae $\Psi_0,\ldots,\Psi_n$ from $\underline{L}$ whose last element Ψ_n is Φ and for $i < n$, Ψ_i is either an element of $\underline{A}$ or a logical axiom or there are $j,k < i$ such that $\Psi_k \triangleq \Psi_j \rightarrow \Psi_i$ (application of modus ponens) or there is a $k < i$ such that $\Psi_i \triangleq \bigwedge_u \Psi_k$ (application of generalization).

We shall write $\underline{A} \vdash \Phi$ for $(\exists S)(PRF_{\langle \underline{L},\underline{A} \rangle}(\Phi,S))$ and say that Φ is syntactically derivable from $\underline{A}$ by means of logical axioms and the rules of inference modus ponens and generalization.

Define $PR(\underline{A}) = \{\Phi; \underline{A} \vdash \Phi\}$, then $\underline{V} \triangleq PR(\underline{A})$ is the set of valid sentences of the elementary theory $\langle \underline{L}, \underline{A} \rangle$.

C) ARITHMETIZATION

The method of arithmetization of the syntax of an elementary theory (id est: first order theory) is due to K. Gödel, hence sometimes called "gödelization". The method consists in adjoining to every formula Φ of an elementary language $\underline{L}_K$ (with alphabet K) a natural number $\gamma(\Phi)$, sometimes written as $\ulcorner \Phi \urcorner$, such that every natural number which is in the range of γ is uniquely "readable" as a formula of $\underline{L}_K$, i.e. has a unique grammatical structure. The definition of γ is by induction on the number of symbols in $\Phi \in \underline{L}_K$ (id est: the complexity of Φ) and is carried out in detail in Feferman [15]. It follows that the sets of gödelnumbers of formulae Φ from At_K and from $\underline{L}_K$ are primitive recursive and similarly the set $\{\gamma(t); t \in T_K\}$ of gödelnumbers of terms is primitive recursive. The proof relation PRF as given above can be put into the form of a number-theoretic function prf and it can be shown that prf is primitive recursive in $\underline{A}$, see Feferman [15] p. 44.

Definition: A first order theory $\underline{T} \triangleq \langle \underline{L}, \underline{V} \rangle$ is axiomatizable iff there is a r.e. subset $\underline{A}$ of $\underline{V}$ such that $\underline{V} \triangleq PR(\underline{A})$. $\underline{A}$ is called a set of axioms for $\underline{T}$. If there is such a finite subset $\underline{A}$, then $\underline{T}$ is called finitely axiomatizable.

W. Craig has proved that if a first order theory $\underline{T}$ has a r.e. set of axioms it has a primitive recursive set of axioms (see his paper: "On axiomatizability within a system", J.S.L. 18 (1953) p.30-32).

Let $\underline{T}_1 \triangleq \langle \underline{L}_1, \underline{V}_1 \rangle$ an $\underline{T}_2 \triangleq \langle \underline{L}_2, \underline{V}_2 \rangle$ be first order theories such that the alphabet K_1 of $\underline{T}_1$ is a subset of the alphabet K_2 of $\underline{T}_2$. Hence $\underline{L}_1 \subseteq \underline{L}_2$. If $\underline{V}_1 \subseteq \underline{V}_2$, then $\underline{T}_2$ is called an extension of $\underline{T}_1$ (in symbols: $\underline{T}_1 \leqslant \underline{T}_2$) and $\underline{T}_1$ is called a subtheory of $\underline{T}_2$.

Definition: Suppose that $\underline{T}_1$ is a subtheory of $\underline{T}_2$. If there is a finite set W_2 of sentences from $\underline{L}_2$ such that $PR(\underline{V}_1 \cup W_2) \triangleq \underline{V}_2$, then $\underline{T}_2$ is called a finite extension of $\underline{T}_1$. If furthermore $K_1 \triangleq K_2$ (equality of the alphabets of $\underline{T}_1$ and of $\underline{T}_2$), then $\underline{T}_2$ is called a <u>finite extension</u> of $\underline{T}_1$ <u>without new constants</u>.

As usually a theory is consistent iff not both Φ and $\neg\Phi$, for some Φ, are derivable from $\underline{T}$. Using the numbertheoretic function prf this can be expressed by a numbertheoretic function **Con** too. We wish to define **Con** for infinite, but r.e. sets $\underline{A}$ of formulae (in this moment we identify formulae Φ with their corresponding gödel-numbers $\gamma(\Phi)$). For this reason it is best to make precise the proof-relation prf too. Hence, let α be the (primitive recursive) function defined on ω such that $\underline{A} \triangleq \{\alpha(n);\ n \in \omega\}$. Now let Prf_α be the number-theoretic formula (with free variables x,y) which expresses the assertion that y is a proof of x on the basis of formulas represented by α (Prf_α is just the predicate corresponding to our previously defined relation $PRF_{\langle \underline{L},\underline{A}\rangle}$). For a rigorous formulation of Prf_α see Feferman [15], or p. 104 in:

[63] R. MONTAGUE: Fraenkel's Addition to the axioms of Zermelo; Fraenkel-Festschrift "Essays on the Foundations of Mathematics", Jerusalem 1966, 2nd Edition, p. 91-114.

Now $\mathbf{Con}_\alpha$ can be defined to be the following formula:

$$\neg \bigvee_x \bigvee_y (Prf\ (x \wedge \neg x, y))\ .$$

$\mathbf{Con}_\alpha$ is a sentence (of Peano-Arithmetic) expressing that the set of formulae represented by α is consistent.

The following theorem is a generalization of Gödel's second underivability theorem (1931) which is due to S. Feferman (Thesis, 1957). Let $\underline{P}$ be the axiomatic system of Peano-Arithmetic (see [15] p. 50). A formula ϕ of $\underline{P}$ defines the set A of natural numbers in $\underline{P}$ iff ϕ has one free variable and for every $n \in \omega$: $\underline{P} \vdash \phi(\underline{n})$ if $n \in A$, and: $P \vdash \neg\phi(\underline{n})$ if $n \notin A$. Here $\underline{n}$ is the nth numeral of $\underline{P}$.

<u>Theorem</u> (Gödel-Feferman): Let A be a consistent set of elementary formulas, and let α be a primitive recursive function which defines A in $\underline{P}$. Then $\underline{P} + \mathbf{Con}_\alpha$ is not relatively interpretable in A (see Feferman [15] p. 90).

D) SYNTACTICAL MODELS

A semantical model (or mostly simply: a model) of a 1st-order theory $\underline{T}$ is given by a (meta-mathematical) set $\underline{J}$ in which some relations are defined such that the set $\underline{V}$ of "valid sentences of $\underline{T}$" are all "true" in $\underline{J}$. In contrast to this a syntactical model of $\underline{T}$ is given by a transformation into another 1st-order theory $\underline{T}^*$. Here is required that the transformed "valid sentences of $\underline{T}$" are all "valid sentences of $\underline{T}^*$". In the case that $\underline{T}$ and $\underline{T}^*$ are axiomatized by $\underline{A}$ and $\underline{A}^*$ respectively this means that the transformed axioms $\underline{A}$ of $\underline{T}$ have to be derivable from $\underline{A}^*$.

The notion of a syntactical model is very old and due to many authors. A. Tarski has called them "Interpretations", since $\underline{T}$ is by the procedure mentioned above "interpreted" in $\underline{T}^*$, see:

[85] A. TARSKI-A. MOSTOWSKI-R.M. ROBINSON: Undecidable theories; (studies in Logic), North Holland Publ.Comp. Amsterdam 1953.

[87] HAO WANG: Arithmetic translations of axiom systems; Transactions of the Amer. Math. Soc. vol. 71 (1951) p. 283-293.

The classical notion of "Interpretation" of one theory in another one has been generalized to "parametrical interpretations" by Petr Hájek:

[29] P. HÁJEK: Syntactic models of Axiomatic Theories; Bull. Acad. Polon. Sci., vol 13 (1965) p. 273-278.

[30] P. HÁJEK: Generalized Interpretability in terms of models - Note to a paper of R. Montague; Casopis pro pěstování matematiki vol. 91 (1966) p. 352-357.

Hájek has defined his generalized notion of Interpretation only for finitely axiomatized theories and used the name "syntactic models" for it. We shall follow Hájek but give the definition for arbitrary r.e. axiomatized theories (the definition can obviously be stated even more generally by writing $\in \underline{V}_2$ instead of $\underline{A}_2 \vdash$ etc.). In order to simplify the notation we assume henceforth that the alphabet K of a first order theory $\underline{T}$ consists of only one sort of variables $v_0, v_1, \dots,$ a sequence of primitive predicates π_i and signs $\neg, \wedge, \bigwedge$ (hence K does not contain any constant or any function letter). This assumption is no loss of generality.

In the following definition let $\underline{T}_1$ and $\underline{T}_2$ be axiomatized 1st order theories such that $\{v_n;\ n \in \omega\}$ is the set of variables of $\underline{T}_1$ and $\{w_n;\ n \in \omega\}$ the set of variables of $\underline{T}_2$ and let I be an index set such that $\{\pi_i;\ i \in I\}$ is the set of primitive predicates of $\underline{T}_1$. n_i will be the number indicating that π_i is n_i-ary.

Definition: Let $\underline{T}_1 \triangleq \langle \underline{L}_1, \underline{A}_1 \rangle$ and $\underline{T}_2 \triangleq \langle \underline{L}_2, \underline{A}_2 \rangle$ be 1st order theories. The set of formulas of $\underline{L}_2$:

$\hat{\Theta}(w_1, \ldots, w_k)$,

$\hat{\Phi}(w_1, \ldots, w_k, w_{k+1})$,

$\hat{\Psi}_i(w_1, \ldots, w_k, \ldots, w_{k+n_i})$, $\quad i \in I$,

is called a translation from $\underline{T}_1$ into $\underline{T}_2$ iff the following three conditions are satisfied:

(1) If w_j is a bound variable in $\hat{\Psi}_i$, then j is odd;

(2) $\underline{A}_2 \vdash \bigvee_{w_1} \ldots \bigvee_{w_k} \hat{\Theta}(w_1, \ldots, w_k)$;

(3) $\underline{A}_2 \vdash \bigwedge_{w_1} \ldots \bigwedge_{w_k} [\hat{\Theta}(w_1, \ldots, w_k) \rightarrow \bigvee_{w_{k+1}} \hat{\Phi}(w_1, \ldots, w_{k+1})]$.

Since $\hat{\Theta}$ characterizes the parameters we shall call $\hat{\Theta}$ the "parameter-condition"; $\hat{\Phi}$ defines (in dependence to the parameters) the objects of the model, hence we shall call $\hat{\Phi}$ the "model-condition". The open formulae $\hat{\Psi}_i$ will be the images (in $\underline{T}_2$) of the primitive predicates π_i (i $\in$ I) of $\underline{T}_1$. Condition (1) is required only in order to avoid confusion of variables in the next definition (otherwise we would have to care about a suitable changement of the bound variables in $\hat{\Psi}_i$.

It seems to be convenient to introduce a replacement operator rep. Therefore let $\tau \triangleq \langle \hat{\Theta}, \hat{\Phi}, \{\hat{\Psi}_i;\ i \in I\} \rangle$ be a translation of $\underline{T}_1$ into $\underline{T}_2$ and let h be the following mapping from the set of free variables $v_0, v_1, v_2, \ldots$ of $\underline{T}_1$ into the set of free variables $w_0, w_1, w_2, \ldots$ of $\underline{T}_2$ given by:

$$h : v_j \mapsto w_{2(j+k)}$$

(k is the number of free variables in $\hat{\Theta}$ and is therefore uniquely determined by τ). The definition of $rep_\tau(\Gamma)$ for $\Gamma \in \underline{L}_1$ is by induction on the length of Γ. Remark that if v_j is free in π_i, then $h(v_j)$ is a free variable of $rep_\tau(\pi_i)$ (this is guaranteed by the curious condition (1) from the definition above).

Definition (of the rep-Operator with respect to a translation τ):

(i) If Γ is an atomic formula of the form $v_{j_1} = v_{j_2}$, then
$\mathrm{rep}_\tau(\Gamma) \triangleq (w_{2(k+j_1)} = w_{2(k+j_2)})$;

(ii) If Γ is an atomic formula of the form
$\pi_i(v_{j_1},\dots,v_{j_{n_i}})$, then
$\mathrm{rep}_\tau(\Gamma) \triangleq \hat{\Psi}_i(w_1,\dots,w_k,w_{2(k+j_1)},\dots,w_{2(k+j_{n_i})})$;

(iii) If Γ is of the form $\Gamma_1 \wedge \Gamma_2$, or $\neg\Gamma_1$, then
$\mathrm{rep}_\tau(\Gamma) \triangleq \mathrm{rep}_\tau(\Gamma_1) \wedge \mathrm{rep}_\tau(\Gamma_2)$,
$\mathrm{rep}_\tau(\Gamma) \triangleq \neg\mathrm{rep}_\tau(\Gamma_1)$ respectively;

(iv) If Γ is of the form $\bigwedge_{v_j} \Gamma^*$, then
$\mathrm{rep}_\tau(\Gamma) \triangleq \bigwedge_{w_{2(k+j)}} [\hat{\Phi}(w_1,\dots,w_k,w_{2(k+j)}) \rightarrow \mathrm{rep}_\tau(\Gamma^*)]$.

The action of rep_τ can be described briefly as follows: in the formula Γ of $\underline{L}_1$ every n_i-ary primitive predicate π_i is replaced by Ψ_i and all quantifiers are restricted to $\hat{\Phi}$. Remark that the image of a sentence Γ under rep_τ is again a sentence if and only if $\hat{\Theta}$ is a sentence, id est: if the translation τ is without parameters.

Definition. A translation $\tau \triangleq \langle \hat{\Theta},\hat{\Phi},\{\hat{\Psi}_i;\ i \in I\}\rangle$ from $\underline{T} \triangleq \langle \underline{L}_1,\underline{A}_1\rangle$ into $\underline{T}_2 \triangleq \langle \underline{L}_2,\underline{A}_2\rangle$ is called a syntactic model of $\underline{T}_1$ in $\underline{T}_2$ iff for every $\Phi \in \underline{A}_1$ the following holds:

$$\underline{A}_2 \vdash \bigwedge_{w_1}\dots\bigwedge_{w_k}[\hat{\Theta}(w_1,\dots,w_k) \rightarrow \mathrm{rep}_\tau(\Phi)]$$

τ is called parametric if $k \neq o$, parameterfree otherwise.

It is easily seen that the notion of a syntactic model (à la Hájek) is a generalization of the classical notion of Interpretation, because, if $J \triangleq \langle \hat{\Phi},\{\hat{\Psi}_i;\ i \in I\}\rangle$ is an interpretation of $\underline{T}_1$ in $\underline{T}_2$ (in the sense of Tarski **[85]**, and Γ any sentence, provable in $\underline{T}_2$, then $\langle \Gamma,\hat{\Phi},\{\hat{\Psi}_i;\ i \in I\}\rangle$ is a (parameterfree) syntactic model of $\underline{T}_1$ in $\underline{T}_2$.

Obviously the following holds: if τ is a translation from $\underline{T}_1$ into $\underline{T}_2$ and $\Psi \in \underline{L}_1$ a logically true formula (i.e. $\emptyset \vdash \Psi$), then

$rep_\tau(\Psi)$ is logically true. This enables to prove by means of the deduction theorem and the finiteness-theorem the following lemma.

Lemma. Let $\tau \triangleq \langle \hat{\theta}, \hat{\Phi}, \{\hat{\Psi}_i;\ i \in I\} \rangle$ be a syntactic model of $\underline{T}_1 \triangleq \langle \underline{L}_1, \underline{A}_1 \rangle$ in $\underline{T}_2 \triangleq \langle \underline{L}_2, \underline{A}_2 \rangle$ and suppose that $\underline{A}_1 \vdash \Phi$.

Then $\underline{A}_2 \vdash \bigwedge_{w_1} \ldots \bigwedge_{w_k} [\hat{\theta}(w_1,\ldots,w_k) \to rep_\tau(\Phi)]$.

The importance of the notion of a syntactic model is contained in the following theorem.

Theorem. Let $\underline{T}_1 \triangleq \langle \underline{L}_1, \underline{A}_1 \rangle$ and $\underline{T}_2 \triangleq \langle \underline{L}_2, \underline{A}_2 \rangle$ be first order theories and suppose that $\underline{T}_2$ is consistent. If there exists a syntactic model τ of $\underline{T}_1$ in $\underline{T}_2$, then $\underline{T}_1$ is consistent too.

Proof. Assume that $\underline{T}_1$ is inconsistent. Then there is a formula Γ of $\underline{L}_1$ such that $\underline{A}_1 \vdash \Gamma \wedge \neg\Gamma$ (and in particular $\underline{A}_1 \vdash \Gamma$). Hence by the preceeding lemma:

(o) $$\underline{A}_2 \vdash \bigwedge_{w_1} \ldots \bigwedge_{w_k} [\hat{\theta}(w_1,\ldots,w_k) \to rep_\tau(\Gamma \wedge \neg\Gamma)], \text{ and}$$

(oo) $$\underline{A}_2 \vdash \bigwedge_{w_1} \ldots \bigwedge_{w_k} [\hat{\theta}(w_1,\ldots,w_k) \to rep_\tau(\Gamma)].$$

By (2) of the definition of "translation" we get from (oo):

(+) $$\underline{A}_2 \vdash \bigvee_{w_1} \ldots \bigvee_{w_k} [\hat{\theta}(w_1,\ldots,w_k) \wedge rep_\tau(\Gamma)].$$

From (o) we get using clause (iii) of the definition of "rep_τ":

$$\underline{A}_2 \vdash \bigwedge_{w_1} \ldots \bigwedge_{w_k} [(\hat{\theta} \vee rep_\tau(\Gamma)) \wedge (\neg\hat{\theta} \vee \neg rep_\tau(\Gamma))].$$

We are interested only in the second member of the conjunction and get:

(++) $$\underline{A}_2 \vdash \neg\bigvee_{w_1} \ldots \bigvee_{w_k} [\theta \wedge rep_\tau(\Gamma)]$$

(+) and (++) show that $\underline{T}_2$ would be inconsistent too, a contradiction to our hypothesis.

Theorem (S. Orey). Let $\underline{T}_1$ and $\underline{T}_2$ be axiomatic 1st order theories, $\underline{T}_2$ be reflexive and $\underline{T}_2$ contain Peano Arithmetic. If for every finite subsystem $\underline{D}$ of $\underline{T}_1$ we have, that $\underline{D}$ has a parameterfree syntactic model in $\underline{T}_2$, then $\underline{T}_1$ has a parameterfree syntactic model in $\underline{T}_2$.

For a proof see Feferman [15] p. 80. A theory $\underline{T}$ is called reflexive iff the consistency of every finite subtheory $\underline{D}$ of $\underline{T}$ can be proved within $\underline{T}$. Further results on syntactic models of set theory will be contained in the following chapters.

E) ZERMELO-FRAENKEL SET THEORY

We call ZF (set theory of Zermelo-Fraenkel) a first order theory with identity whose alphabet, formulae and axioms are defined as follows:

The Alphabet of ZF: One sort of variables $x_0, x_1, \dots, x_n, \dots (n \in \omega)$ ($x, y, z, \dots$ stand for these variables), a binary predicate ε, and logical symbols: $\neg, \vee, \bigvee$, = and brackets.

The Formulae of ZF: $x \varepsilon y$ and $x = y$ are (atomic) formulae, and if Φ and Ψ are formulae, then $\neg(\Phi)$, $(\Phi) \vee (\Psi)$ and $\bigvee_x(\Phi)$ are formulae.

We follow the usual conventions which allow us to omit brackets in some cases (see Hilbert-Ackermann p. 74).

The axioms of ZF:

(0) Null-set: $\bigvee_x \bigwedge_y [\neg y \varepsilon x]$.

(I) Extensionality (Axiom der Bestimmtheit):
$\bigwedge_x \bigwedge_y [\bigwedge_z (z \varepsilon x \leftrightarrow z \varepsilon y) \rightarrow x = y]$,

(II) Axiom of pairs:
$\bigwedge_x \bigwedge_y \bigvee_z [\bigwedge_u (u \varepsilon z \leftrightarrow u = x \vee u = y)]$,

(III) Axiom of unions (Sums):
$\bigwedge_x \bigvee_y \bigwedge_z [z \varepsilon y \leftrightarrow \bigvee_u z \varepsilon u \varepsilon x]$,

(IV) Axiom of infinity:
$\bigvee_x \bigvee_y [y \varepsilon x \wedge \bigwedge_z (z \varepsilon x \rightarrow \bigvee_u (u \varepsilon x \wedge u \neq z \wedge \bigwedge_v (v \varepsilon z \rightarrow v \varepsilon u)))$,

(V) Power-set (Potenzmengen-Axiom):
$\bigwedge_x \bigvee_y \bigwedge_z [z \varepsilon y \leftrightarrow \bigwedge_u (u \varepsilon z \rightarrow u \varepsilon x)]$,

(VI_Φ) Axioms of substitution (Ersetzungsaxiome)

$$\bigwedge_x \overset{1}{\bigvee}_y \Phi(x,y) \rightarrow \bigwedge_a \bigvee_b \bigwedge_y [y\varepsilon b \leftrightarrow \bigvee_x (x\varepsilon a \wedge \Phi(x,y))].$$

(VII) Axiom of regularity (Fundierungsaxiom)

$$\bigwedge_x [\bigvee_y y\varepsilon x \rightarrow \bigvee_y (y\varepsilon x \wedge \bigwedge_z (\neg z\varepsilon y \vee \neg z\varepsilon x))].$$

In the schema (VI_Φ) the formula Φ is supposed not to contain any free occurence of the variable b. (VI_Φ) implies the axioms of subsets (Aussonderungsaxiome).

The system of axioms (0) - (VII) is called ZF. The subsystem consisting of axioms (0) - (VI_Φ) is called ZF°. Hence

$$ZF \triangleq ZF^\circ + \text{"Axiom of regularity"}.$$

Sets, totalities and classes have always played an important rôle in mathematics though this has not been recognized explicitly It was Georg Cantor (1845-1918) who has formulated explicitly the ever used properties of sets. According to the methodology of mathematics he investigated the interrelations and interdependencies between these various properties. This lead to the socalled Set Theory.

[8] G. CANTOR: Gesammelte Abhandlungen (Edited by E. Zermelo), Springer-Verlag Berlin 1932, Reprinted 1962 by G. Olms, Hildesheim.

Cantor's Set Theory can be viewed as a consequent attempt to transfer our comtemplations with respect to the domain of finite sets to domains including infinite sets.
Exempla gratia: the principle, that the image of a finite set is again a finite set, yields the axiom of replacement (VI_Φ). On the other hand there are principles which cannot be translated, such as $m < m + 1$ for finite cardinals m. Hence it is not too much surprising that between these extremes there is a large class of assumptions for which it is hard to saywhether they are true only when restricted to finite sets. A very classical example for this situation is the axiom of choice:

$$\text{(AC)} \quad \bigwedge_x (\bigwedge_y (y\varepsilon x \rightarrow y \neq \emptyset) \rightarrow \bigvee_f (Fnc(f) \wedge \bigwedge_{y\varepsilon x} f(y)\varepsilon y))$$

There are other problems which do not have an analogue in the domain of finite sets, such as the continuum-problem. But as in the case of the axiom of choice (AC) our questions are wether the Cantorian principles, which are abstracted from the domain of finite sets, are sufficent to decide them.

The original Cantorian set Theory [as based on the "Ideal Kalkül", whose axioms are extensionality and comprehension $\bigvee_x(\bigwedge_y y\varepsilon x \leftrightarrow \Phi(x))]$ was contradictory (take Russell's predicate $\neg x\varepsilon x$). Since 1908 many axiomatic systems for set theory have been given which seem not to give raise to the Russell-antinomie. Roughly they can be divided into two groups. The first group are type-theoretic foundational systems, such as the Principia Mathematica PM of Russell-Whitehead, the systems NF ("New Foundations") and ML ("Mathematical Logic") of Quine, the systems of N. da Costa (Indagationes Math. 27 (1965)), J. Houdebine (Thèse, Rennes 1967). The second group are Extensions of Zermelo's set theory (1908):

[88] E. ZERMELO: Untersuchungen über die Grundlagen der Mengenlehre; Math. Annalen 65(1908) p.261-281.

[89] E. ZERMELO: Über den Begriff der Definitheit in der Axiomatik; Fund. Math. 14(1929) p.339-344.

such as Johann von Neumann's System (Math. Zeitschrift 27(1928) p.669-752, and J. für die reine u. angew. Math.(Crelle)154(1925) p.219-240), the system of Bernays (contained in a sequence of 7 papers in the J.S.L.,vol.2,6,7,8,13,19, a modified version is contained in his book, Amsterdam 1958), the version in Gödel's monograph (Princeton 1940), the ZF-version of Bourbaki, Skolem, Thiele, Sonner and others. One of the nicest foundational system has been given by:

[1] W. ACKERMANN: Zur Axiomatik der Mengenlehre; Math. Ann.131 (1956) p.336-345.

Since we will not discuss this system, we refer the reader to the following articles:

[45] A. LÉVY: On Ackermann's set theory; JSL 24(1959) p.154-166.

[56] A. LÉVY - R.L. VAUGHT: Principles of partial reflection in the set theories of Zermelo and Ackermann; Pacific J. Math. 11(1961) p.1045-1062.

[27] R. GREWE: On Ackermann's set theory; Doctoral Dissertation, University of California, Los Angeles 1966.

[28] R. GREWE: Natural models of Ackermann's Set Theory; JSL 34(1969) p.481-488.

[71] W.N. REINHARDT: Topics in the Metamathematics of Set Theory Thesis, University of Wisconsin 1967? See also Reinhardt's Abstract: "Ackermann's Set Theory Coincides with ZF" in the AMS-Notices, vol.13(1966) p.727.

Zermelo's original system Z contained only the axioms (O) - (V) plus the "Aussonderungsschema":

$$\bigwedge_x \bigvee_y [\bigwedge_z (z \varepsilon y \leftrightarrow z \varepsilon x \wedge \Phi(z))].$$

The "Fundierungsaxiom" and the "Ersetzungsaxiom" (VI_Φ) are introduced by A. Fraenkel (in his article: Zu den Grundlagen der Cantor-Zermeloschen Mengenlehre, Math. Annalen 86(1922) p.230-237). Zermelo has adopted the whole system (O) - (VII) in his article on "Grenzzahlen und Mengenbereiche" (Fund. Math. 16(1930) p.29-47) and has called it ZF.

We assume that the reader is familiar with the classical development of ZF-set theory. There are many textbooks on this subject, such as J. Rubin: Set theory for the mathematician (Holden-Day 1967) or: K. Kuratowski - A. Mostowski: Set Theory (North-Holland Publ. Comp. Amsterdam 1968).
We shall use the following conventions:
For sets $x,y,\underline{P}(x)$ is the power-set of x, $x \cap y$ the intersection, $x \cup y$ the union, $x - y$ the difference and $x \times y$ is the cartesian product. Ordered pairs $\langle x,y \rangle$ are defined à la Kuratowski: $\langle x,y \rangle = \{\{x\},\{x,y\}\}$. If s is a set of ordered pairs than $pr_1(s)$ is the projection of s to the first coordinate: $pr_1(s) = \{x; \bigvee_y \langle x,y \rangle \varepsilon s\}$. $pr_2(s)$ is similarly defined as projection on the second coordinate. $Fnc(f)$ is the open formula expressing that f is a function. $pr_1(f)$ is the domain of f and $pr_2(f)$ the range of f, sometimes written as $Dom(f)$, $Rg(f)$ respectively. We shall freely use abstraction terms $\{x; \Phi(x)\}$ for ZF-formulae Φ. This yields a definitional extension of ZF. In such an extension of ZF the set of logical axioms is enriched by Church's conversion principle:

$$x \varepsilon \{y; \Phi(y)\} \leftrightarrow \Phi(y/x)$$

where y is the free variable of Φ and $\Phi(y/x)$ is the formula obtained from Φ by substituting x for y at all places of free occurrences of y in Φ. It is supposed that x is free for y in Φ.

Let α and β denote Ordinalnumbers (à la v. Neumann). For any set x define $R_\alpha(x) = \bigcup\{\underline{P}(R_\beta(x));\ \beta < \alpha\}$. For $x = \emptyset$ one gets the well-known v. Neumann's "Stufen" and we shall write $V_\alpha = R_\alpha(\emptyset)$.
The Mirimanoff-rank $\rho(x)$ of a set is defined to be the first α such that $x \in V_{\alpha+1}$. The "Stufen" V_α and the rank-function ρ have the following properties:

<u>Lemma</u>: 1) $\alpha \leqslant \beta \leftrightarrow V_\alpha \subseteq V_\beta$,

2) $\rho(\alpha) = \alpha$,

3) $\rho(x) = \bigcup\{\rho(y) + 1;\ y \in x\}$,

4) If $x \in y \in V_\alpha$ or $x \subseteq y \in V_\alpha$, then $x \in V_\alpha$,

5) $x \in V_\alpha \leftrightarrow \rho(x) < \alpha$,

6) $V_{\alpha+1} = P(V_\alpha)$,

7) $V_\lambda = \bigcup\{V_\alpha;\ \alpha < \lambda\}$ if λ is a limit ordinal,

8) If A is any set of ordinals, then $\bigcup_{\alpha \in A} V_\alpha = V_\beta$ for some ordinal β.

9) $x \subseteq V_{\rho(x)}$,

10) $x \in V_{\rho(x)+1}$.

Proof by induction.

Note that $V = \bigcup\{V_\alpha;\ \alpha \text{ an ordinal}\}$, where V is the class of all sets, is in ZF^0 equivalent to the axiom of regularity.
The class $\bigcup V_\alpha$ can be used to construct a parameterfree*) ZF in ZF^0 . This shows the relative consistency of ZF with respect to ZF^0 (see the lemma in section D).

F) THE PRINCIPLE OF REFLECTION

For a given formula Φ of the ZF-language and a class-term C let $Rel(C,\Phi)$ be the formula obtained from Φ by restricting every quantifier in Φ to C , id est. by replacing each occurrence of $\bigwedge_x \psi$ or $\bigvee_x \psi$ by $\bigwedge_x (x \in C \rightarrow \psi)$ or $\bigvee_x (x \in C \wedge \Psi)$, respectively (alphabetic change of bound variables may also be needed). The rigorous

*) insert: syntactic model of.

definition (by induction on the length of Φ) of $\mathrm{Rel}(C,\Phi)$ can be given analoguously to our definition of $\mathrm{rep}_\tau(\Phi)$ in section D.

Definition: Let Φ and C be given. Φ <u>is convenient for</u> C iff the following is provable in ZF:

$$(*)\quad \bigwedge_{x_1 \in C} \cdots \bigwedge_{x_n \in C}[\Phi(x_1,\ldots,x_n) \leftrightarrow \mathrm{Rel}(C,\Phi)]$$

A partial ordering $\trianglelefteq$ between ZF-formulae in prénexform is defined by $\phi_1 \trianglelefteq \phi_2$ iff there are finitely many quantifiers $\bigvee x_1,\ldots,\bigvee x_k$ such that ϕ_2 is obtained from ϕ_1 by putting in front of ϕ_1 negation-symbols $\neg$ and these quantifiers $\bigvee x_i$ in a certain order. It is known that every formula is equivalent to a formula (with the same set of free variables) in prenex form.

Definition: Φ in prenex form is hereditarily convenient for C iff $(*)$ holds not only for ϕ but also for every $\psi \trianglelefteq \phi$.

Lemma: Let ϕ be a ZF-formula in prenex form and let $\{X_i;\ i \in \omega\}$ be a nested sequence of sets (i.e. $X_i \subseteq X_{i+1}$). If ϕ is hereditarily convenient for every X_n, ϕ is hereditarily convenient for $X = \bigcup_{i\in\omega} X_i$ too.

Proof (by induction on the length of ϕ): If ϕ does not have quantifiers, then the lemma is immediate. The case where ϕ has the form $\neg\psi$ is obvious. Now suppose ϕ has the form $\bigvee_x \psi(x,x_1,\ldots,x_n)$. By assumption ϕ is hereditarily convenient for each X_n. Hence the same holds for ψ. By induction hypothesis ψ is hereditarily convenient for X. Now $(*)$ follows immediately for $\bigvee_x \psi$ and the lemma is proved.

Theorem: (R.Montague - A.Lévy): Every instance of the following schema of complete reflexion is a theorem of ZF:

$$(\mathrm{CR}^*)\qquad \bigwedge_\alpha \bigvee_\beta [\mathrm{Lim}(\beta) \wedge \alpha < \beta \wedge \bigwedge_{x_1 \in V_\beta} \cdots \bigwedge_{x_n \in V_\beta} [\Phi \leftrightarrow \mathrm{Rel}(V_\beta,\Phi)]]$$

Here α and β are Variables ranging over ordinals and $\mathrm{Lim}(\beta)$ expresses that β is a limit ordinal, i.e. $\beta \neq 0 \wedge \beta = \bigcup\beta$. It is understood that $x_1,\ldots,x_n$ are precisely the free variables of Φ.

Proof: It is sufficient to prove the theorem only for the case that ϕ is in prenex form. If ϕ is without quantifiers, the theorem is obviously true for $\beta = \alpha + \omega$. The case $\phi \triangleq \neg\psi$ is also very trival. Hence let us suppose that $\phi(x_1,\ldots,x_n)$ has the form $\bigvee_x \psi(x,x_1,\ldots,x_n)$. By the induction hypothesis there is for every α a limit-ordinal β such that ψ is hereditarily convenient for V_β. Let $\Gamma(y,x_1,\ldots,x_n)$ be the formula saying that y is the set of sets x of minimal rank such that $\psi(x,x_1,\ldots,x_n)$. Define a countable sequence of ordinals $\beta_0 < \beta_1 \ldots < \beta_n < \ldots$ in the following way:

Let β_0 be the first ordinal $> \alpha$ such that ψ is hereditarily convenient for V_{β_0}. Having defined β_i for $i \leqslant 2n$, put

$$\beta_{2n+1} = \mathrm{Min}\{\gamma;\ \beta_{2n} < \gamma \wedge \bigwedge_{x_1 \varepsilon V_{\beta_{2n}}} \ldots \bigwedge_{x_n \varepsilon V_{\beta_{2n}}} [\Gamma(y,x_1,\ldots,x_n) \rightarrow y \varepsilon V\gamma]\}.$$

Now let β_{2n+2} be the least ordinal $> \beta_{2n+1}$ such that ψ is hereditarily convenient for $V_{\beta_{2n+2}}$. Obviously $\beta = \lim_{n\varepsilon\omega} \beta_n$ is a limit ordinal and

$$V_\beta = \bigcup_{n\varepsilon\omega} V_{\beta_n} = \bigcup_{n\varepsilon\omega} V_{\beta_{2n}}.$$

(see the lemma in section E). In order to show that V_β is convenient for $\bigvee_x \psi(x,x_1,\ldots,x_n)$, let $a_1,\ldots,a_n \varepsilon V_\beta$ be given. Hence the a_i's are elements of partial universes V_{β_i} and let these finitely many V_{β_i}'s be contained in $V_{\beta_{2k}}$. If $\phi(a_1,\ldots,a_n)$ holds, then $\bigvee_x \psi(x,a_1,\ldots a_n)$ holds too. Hence

$$\bigvee_x [\bigvee_y \, x \varepsilon y \wedge \Gamma(y,a_1,\ldots,a_n) \wedge \psi(x,a_1,\ldots,a_n)]$$

holds too. It follows from the definition of β_{2k+1}, that there is an object a in $V_{\beta_{2n+2}}$ such that $\psi(a,a_1,\ldots,a_n)$ holds. Since $V_{\beta_{2n+2}} \subseteq V_\beta$, $a \varepsilon V_\beta$, it follows from the induction hypothesis, that $\mathrm{Rel}(V_\beta,\psi(a))$ holds (remark that by the construction $\psi(a)$ is hereditarily convenient for V_{2n}, hence by the previous lemma for V_β too). Hence $\mathrm{Rel}(V_\beta,\phi)$. The direction $\mathrm{Rel}(V_\beta,\phi) \rightarrow \phi$, when the free variables are restricted to range only over V_β, is obvious, because if $\mathrm{Rel}(V_\beta,\Phi)$ holds in V_β, then there is an object $a \varepsilon V_\beta$ such that $\mathrm{Rel}(V_\beta,\Psi(a))$ holds in V_β. By the induction hypothesis $\Psi(a,a_1,\ldots a_n)$ holds for every sequence $\langle a_1,\ldots,a_n\rangle \varepsilon V_\beta^n$, with $a \varepsilon V_\beta$. Hence $\bigvee_x \psi(x,a_1,\ldots,a_n)$ holds too for every sequence $\langle a_1,\ldots,a_n\rangle \varepsilon V_\beta^n$, q.e.d.

The schema (CR*) is called a schema of (complete) reflexion because it follows that for any closed formula Φ, if Φ holds in the whole universe V, Φ must hold in some partial universe V_β. Since the converse holds too, the reflection principle is called complete. By (CR*) a property Φ of V is "reflected" to be already a property in some V_β.

Let S be the theory based upon extensionality, Aussonderung, pairing, sum set, power set and the following schema of regularity:

(VII$^*_\Phi$) $\bigvee_x \Phi(x) \rightarrow \bigvee_y [\Phi(y) \wedge \bigwedge_z (z \varepsilon y \rightarrow \neg \Phi(z))]$.

We want to show, that ZF = S + Infinity + Replacement and that ZF = S + (CR*). Our axiom of regularity (VII) is the socalled "set-form". In the V.Neumann-Bernays-Gödel set theory (NBG, see Gödel's monograph), one can state regularity also in a "class-form": $\bigwedge_X (X \neq \emptyset \rightarrow \bigvee_x (x \varepsilon X \wedge x \cap X = \emptyset))$. Gödel has pointed out (see Bernays, J.S.L. 13(1948) p.68) that the set form and the class-form of the axiom of regularity are equivalent on the basis of the axioms of groups A,B,C. Since ZF does not have class-Variables, the appropriate "class-form" of the axiom of regularity can be stated in ZF by the schema (VII$^*_\Phi$) above. Obviously (VII$^*_\Phi$) implies (VII) (on the basis of the remaing ZF-axioms), since for a given set u take $x \varepsilon u$ as Φ in (VII$^*_\Phi$). The next theorem shows, that also the converse is true. We take the proof from the following paper:

[36] K. HAUSCHILD: Bemerkungen, das Fundierungsaxiom betreffend; Zeitschrift f. math. Logik u. Grundl. d. Math. 12 (1966) p.51-56.

<u>Theorem</u>: Every instance of (VII$^*_\Phi$) is a theorem of ZF. Hence the set-form of the) axiom of foundation (VII) and the axiom schema of foundation (VII$^*_\Phi$) are equivalent in ZF^0.

<u>Proof</u>: Let u be any set such that $\Phi(u)$ holds. From the axioms of replacement and infinity it follows, that every set x has a transitive closure, i.e. there is a set y, such that $x \subseteq y$ and $\bigwedge_z \bigwedge_w (z \varepsilon w \wedge w \varepsilon y \rightarrow z \varepsilon y)$. $Cl(x) = \{x\} \cup x \cup \bigcup x \cup \bigcup\bigcup x \cup \ldots$ is the transitive closure of x. Since $u \varepsilon Cl(u)$ and $\Phi(u)$ hold, we have $\bigvee_x (x \varepsilon Cl(u) \wedge \Phi(x))$. Now let $\psi(x,y)$ be the following formula $\Phi(x) \wedge x = y$. Hence the replacement axiom for $\psi(x,y)$ yields a set t such that $x \varepsilon t \leftrightarrow (x \varepsilon Cl(u) \wedge \Phi(x))$. By the set form of the axiom of regularity (VII) t has an element v such that $z \varepsilon v \rightarrow \neg z \varepsilon t$. Now, let w be an arbitrary element of v. Since $\neg w \varepsilon t$ we have that either $\neg w \varepsilon Cl(u)$ or $\neg \Phi(w)$. But $w \varepsilon v$, $v \varepsilon t$, $t \subseteq Cl(u)$ and Cl(u) is transitive, hence $w \varepsilon Cl(u)$. Therefore it must hold that $\neg \Phi(w)$. Furthermore $\Phi(v)$, since $v \varepsilon t$. Hence we have obtained a set v such that $\Phi(v)$ and $\neg \Phi(w)$ for every element w of v. Hence the instance of the schema of foundation for Φ is proved', q.e.d.

Let ZF* be the system ZF without the axiom of infinity. K.Hauschild has shown in **[36]** that (VII$^*_\Phi$) is not provable for every Φ in ZF*. R.B.Jensen-M.E.Schröder have shown (in: Archiv math. Logik 12(1969) p.119-133) that the schema (VII$^*_\Phi$) is not provable in Z + VII).

Now we shall show that the schema of complete reflection is within the frame of S (= general set theory) equivalent to the conjunction of the axiom of infinity and replacement. Since we are not able to show within S that for given α, V_α exist (the axiom of replacement would be needed), we have to take the following different formulation:

(CR) Schema of complete reflection: Let Φ be a ZF-formula with free variables $x_1,\dots,x_n$:

$$\bigwedge_y \bigvee_z [y\varepsilon z \wedge \mathrm{Smc}^S(z) \wedge \bigwedge_{x_1 \varepsilon z}\dots\bigwedge_{x_n \varepsilon z}(\Phi \leftrightarrow \mathrm{Rel}(z,\Phi))].$$

Here $\mathrm{Smc}^S(z)$ is the ZF-formula saying that z is a standardcomplete (i.e. transitive) model of general set theory S, more precisely, $\mathrm{Smc}^S(z)$ is the formula:

$$\bigwedge_x \bigwedge_y (x\varepsilon y \varepsilon z \to x\varepsilon z) \wedge \bigwedge_x \bigwedge_y (x\varepsilon z \wedge y\varepsilon z \to \{x,y\} \varepsilon z) \wedge \bigwedge_x (x\varepsilon z \to P(x)\varepsilon z \wedge \bigcup x\varepsilon z).$$

Here P(x) is the power set of x. Since $x \subseteq y\varepsilon z \to x\varepsilon P(y)\varepsilon z \to x\varepsilon z$, z is a model of the Aussonderungsaxiom, if $\mathrm{Smc}^S(z)$. In ZF it is provable that

$$\bigwedge_z [\mathrm{Smc}^S(z) \leftrightarrow \bigvee_\alpha (\mathrm{Lim}(\alpha) \wedge z = V_\alpha)],$$

hence (CR) and (CR') are equivalent in ZF. In order to show that ZF = S + (CR), we have to verify first, that in S, (CR) is equivalent to:

$$(\mathrm{CR}_m) : \bigwedge_y \bigvee_z [y\varepsilon z \wedge \mathrm{Smc}^S(z) \wedge \bigwedge_{x_1 \varepsilon z}\dots\bigwedge_{x_n \varepsilon z} \bigwedge\!\!\!\bigwedge_{i=1}^{m} (\Phi_i \leftrightarrow \mathrm{Rel}(z,\Phi_i))],$$

where $\bigwedge\!\!\!\bigwedge$ means conjunction and Φ_i $(1 \leq i \leq m)$ are ZF-formulae whose free variables are at most $x_1,\dots,x_n$. The implication (CR) $\to$ (CR_m) is proved by means of the following idea: the enumeration $i \mapsto \Phi_i$ $(i \leq m)$ can be implemented within the ZF-language by defining Ψ to be the following formula ($\bigvee\!\!\!\bigvee$ stands for disjunction):

$$\bigvee\!\!\!\bigvee_{i=1}^{m} (t = \underline{i} \wedge \Phi_i).$$

Since the natural numbers are absolute with respect to standard complete models, we have:

$$\bigwedge_z [\mathrm{Smc}^S(z) \to (\mathrm{Rel}(z,\Psi) \leftrightarrow \bigvee\!\!\!\bigvee_{i=1}^{m} (t = \underline{i} \wedge \mathrm{Rel}(z,\Phi_i)))].$$

$\underline{i}$ is the ith-numeral. Since $\mathrm{Smc}^S(z) \to \omega \subseteq z$ we may substitute $\underline{j}$ for t in the reflection formula given by (CR) with respect to Ψ. Thus we have proved (CR_m).

The equivalence S $\vdash$ (CR) $\leftrightarrow$ (CR_m) is due to A.Lévy, contained in his paper (p.229):

[46] A. LÉVY: Axiom schemata of strong infinity in axiomatic set theory; Pacific J. Math. 10(1960)p.223-238.

Theorem (Lévy [46]): ZF = S + (CR).

Proof: In order to show that S + (CR) $\subseteq$ ZF we need to show only that every instance of the axiom schema of Aussonderung is a theorem of ZF. Hence let $\Phi(x)$ be a ZF-formula. If ZF $\vdash \neg \bigvee_x \Phi(x)$, then the empty set $\emptyset$ satisfies the Aussonderungsaxiom for Φ. If ZF $\vdash \bigvee_x \Phi(x)$, then the replacementaxiom with respect to $\psi(x,y) \triangleq \Phi(x) \wedge x = y$ gives a set b which satisfies the Aussonderungsaxiom for Φ.

We show that ZF $\subseteq$ S + (CR). The axiom of Null-set follows directly from the Aussonderungsaxiom with $x \neq x$ as $\Phi(x)$. Infinity is obvious since every set z with $Smc^S(z)$ has the properies required by the axiom of infinity. Now we prove the replacement axiom: Let $\Phi(v,w)$ be given with $v, w, x_1, \dots, x_n$ as the only free variables. Let $\Gamma(x)$ denote the formula:

$$\bigwedge_r \bigwedge_s \bigwedge_t [(\Phi(r,s) \wedge \Phi(r,t) \to s = t) \to \bigvee_y \bigwedge_w [w \varepsilon y \leftrightarrow \bigvee_v (v \varepsilon x \wedge \Phi(v,w))]] .$$

By (CR), or equivalently (CR_4), there is a set z such that $Smc^S(z)$ and for all $x_1, \dots, x_n, x, v, w \varepsilon z$:

$$[\Phi \leftrightarrow Rel(z,\Phi)] \wedge [\bigvee_w \Phi \leftrightarrow Rel(z, \bigvee_w \Phi)] \wedge [\Gamma \leftrightarrow Rel(z,\Gamma)]$$
$$\wedge [\bigwedge_{x_1} \dots \bigwedge_{x_n} \bigwedge_x \Gamma \leftrightarrow Rel(z, \bigwedge_{x_1} \dots \bigwedge_{x_n} \bigwedge_x \Gamma) .$$

The formula in the first brackets says that Φ is absolut with respect to z, hence, by the second formula Φ maps members of z on members of z. By the Aussonderungsaxiom Φ maps the members $x \varepsilon z$ (hence $x \subseteq z$, since z is transitiv) onto a subset y of z. Thus, if $x_1, \dots, x_n, x \varepsilon z$, then Γ. Hence by the third formula: $Rel(z,\Gamma)$. But the universal closure of $Rel(z,\Gamma)$ is $Rel(z, \bigwedge_{x_1} \dots \bigwedge_{x_n} \bigwedge_x \Gamma)$ thus the fourth formula gives us Γ which is the replacement axiom with respect to Φ, q.e.d.

Corollary: In the set theory S the axiom schema of replacement in conjunction with the axiom of infinity is equivalent to the schema of complete reflection (CR).

Let $\mathcal{T}_1$ and $\mathcal{T}_2$ be 1st order theories with the same language and $\mathcal{T}_1$ axiomatizable, $\mathcal{T}_2$ an extension of $\mathcal{T}_1$ containing Peano-Arithmetic, then $\mathcal{T}_2$ is called essentially reflexive over $\mathcal{T}_1$ iff:

$$\mathcal{T}_2 \vdash \Phi \rightarrow \mathrm{Con}(\mathcal{T}_1 + \Phi)$$

for every sentence Φ of the language of $\mathcal{T}_1$. R.Montague [63] has shown that ZF is essentially reflexive over Zermelo's Aussonderungs-schema. Furthermore, he has proved that no consistent extension (without new constants) of ZF is a finite extension of Zermelo set theory Z. A slightly weaker result was obtained by K.Hauschild ("Ein Beitrag zur Metatheorie der Mengenlehre", Zeitschr. f. math. Logik u. Grundl. d. Math.9(1963)p.291-314). Hauschild has shown:

"The system ZF and every extension (without new constants) of ZF which has wellordered models is not finitely axiomatizable".

Notice, that in contrast to ZF the NBG-set theory is (as Bernays has shown) finitely axiomatizable. - Here we shall prove only the following result which states that ZF is not a finite extension of Zermelo-set theory Z.

Lemma. If $u \subseteq w$, then $\mathrm{Rel}(u,\Phi) \leftrightarrow \mathrm{Rel}(w,\mathrm{Rel}(u,\Phi))$.

Proof by induction on the length of the ZF-formula Φ.

Theorem (R.Montague): ZF-set theory is not a finite extension of Zermelo's set theory Z.

Proof. Suppose that ZF is consistent and a finite extension of Z, id est: $ZF = Z + \Phi_1 + \ldots + \Phi_n$ for $\Phi_1, \ldots, \Phi_n$ $(n < \omega)$ sentences of ZF-language. Let Φ be the conjunction of $\Phi_1, \ldots, \Phi_n$. By the assumption Φ is consistent with Z. Since $ZF = Z + \Phi$, Φ holds in the whole universe V. By (CR*) there are partial universes V_β, $\beta > \omega$, such that Φ is true in V_β. Let α be the least such ordinal $(> \omega)$. By the preceeding lemma V_α is convenient for $\mathrm{Rel}(V_\gamma,\Phi)$ (The formula $\mathrm{Rel}(V_\gamma,\Phi)$ has one free variable γ). Now let ψ be the following ZF-formula:

$$\bigvee_\gamma [\mathrm{Lim}(\gamma) \wedge \omega < \gamma \wedge \mathrm{Rel}(V_\gamma,\Phi)]$$

Then $\mathrm{Rel}(V_\alpha,\psi)$ is the following formula:

$$\bigvee_\gamma [\gamma \in V_\alpha \wedge \mathrm{Lim}(\gamma) \wedge \omega < \gamma \wedge \mathrm{Rel}(V_\gamma,\Phi)].$$

Since $\gamma \in V_\alpha$ implies $\gamma < \alpha$ and α was minimal we infer that $\mathrm{Rel}(V_\alpha,\psi)$ is not true and hence ψ is not true in V_α. On the other hand V_α is a model of Z (since $\mathrm{Lim}(\alpha)$) and, by the choice of α, also a model of Φ. Since $ZF \vdash \psi$ (via (CR*)), $Z + \Phi$ cannot give the whole ZF-theory, q.e.d.

<u>1. Corollary</u>: ZF is not finitely axiomatizable.
<u>2. Corollary</u>: ZF is reflexive (i.e. the consistency of every finite subtheory of ZF can be proved within ZF).

The first corollary was obtained (by different methods) also by K.Hauschild [36]. A.Lévy has generalized these corollaries and in particular the results of R.Montague in his papers [46] and

[47] A.LÉVY: On a spectrum of set theories; Illinois J.Math.4 (1960)p.413-424.

In [46] Lévy has extended Montague's results to a transfinite Hierarchy of set theories:

$$Z, ZF, ZM_1, ZM_2, \ldots, ZM_\alpha, \ldots$$

where $ZM_{\alpha+1}$ results from ZM_α by adding suitable (strong) reflection principles. In [47] Lévy constructs further Hierarchies between any two consecutive theories ZM_α, $ZM_{\alpha+1}$. The construction is carried out between ZF and ZM:

$$ZF^1, ZF^2, \ldots, ZF^{\beta+1}, ZF^{\beta+2}, \ldots$$

for every ZF-definite ordinal β. Here $ZF^{\beta+1}$ is roughly the set of all sentences ψ which are true in natural models $\langle V_\theta, \in\rangle$, where θ is an inaccessible number such that V_θ contains at least β many inaccessibles, i.e. $\beta + 1$ many ZF-universa. Lévy shows that ZF^1 is essentially reflexive over ZF and that $ZF^{\beta+2}$ is essentially reflexive over $ZF^{\beta+1}$.

In another paper of A.Lévy: "Comparision of Subtheories" (Proc. A.M.S. 9(1959)p.942-945, Errata vol.10) it is shown that the numbertheory of ZF is an essentially infinite extension of the numbertheory of Z and that the same holds for ZF + "There are inaccessible cardinals" and ZF. More precisely let τ be the usual interpretation (parameterfree syntactic model) of Peano Arithmetic $\mathcal{P}$ in set theory. Then let $Z/\mathcal{P} \triangleq \{\Phi$ formula of Peano-Arithmetic; $Z \vdash rep_\tau(\Phi)\}$ and $ZF/\mathcal{P}$ similarily defined. Then: $ZF/\mathcal{P}$ is an essentially infinite extension of $Z/\mathcal{P}$ and $(ZF + \bigvee_\theta \theta$ inaccessible$)/\mathcal{P}$ is an essentially infinite extension of $ZF/\mathcal{P}$. These results depend essentially on the existence of certain truth definitions with some particular properties.

There are interesting finitely axiomatizable subtheories of ZF. If we admit in the schema of Aussonderung

$$\bigwedge_{p_1} \ldots \bigwedge_{p_n} \bigwedge_y \bigvee_x \bigwedge_t [t \in x \leftrightarrow (t \in y \wedge \Phi(t, p_1, \ldots, p_n))]$$

only those formulas Φ in which occur solely limited quantifiers, i.e.quantifiers having the form $\bigwedge_x (x \in y \rightarrow \ldots)$ as well as $\bigvee_x (x \in y \wedge \ldots)$, then we get a weaker system than Z. The set of Σ_0-Aussonderungsaxioms (in Lévy's terminology) as just described is finitely axiomatizable. Mostowski credits this result to Tarski (in: "The present state ..." p.20), but a proof has never been published. A generalization of this result:"every set of axioms of ZF which are Σ_n or Π_n-formulas for some $n \leq k$, where k is any fixed natural number , is finitely axiomatizable" is due to E.J.Thiele (the finite axiomatization is essentially given by the Bernays-Gödel operations). See the paper:

[84] E.J.THIELE: Über endlich axiomatisierbare Teilsysteme der ZF-Mengenlehre; Zeitsch. f. math. Logik u. Grundlagen d. Math. 14(1968)p.39-58.

CHAPTER II

Constructible Sets

In this chapter we discuss the axiom of choice (AC), the generalized continuum hypothesis (GCH) and Gödel's axiom of constructibility V = L and present Gödel's proof of the relative consistency of these three statements with the axioms of ZF-set theory. We conclude the chapter with a proof of some results of H.Putnam concerning the hierarchy M_α of constructible sets.

A) THE AXIOM OF CHOICE AND THE CONTINUUM HYPOTHESIS

Mathematicians of the 19th century have very often and without any skruple used the hypothesis that for any infinite collection of nonempty sets we are able to choose from every member of the collection one element. With respect to 19th century mathematics it is merely a philosophical question whether human beings are able to arrange infinite choice sequences in their short live; It was G.Peano (in an article on integrability of ordinary differential equations, Math. Annalen 37(1890) p.182-229, see p.210) who first perceived the doubtfulness of such a hypothesis. He wrote: "Mais on ne peut pas appliquer une infinité de fois une loi arbitraire avec laquelle à une classe on a fait correspondre une individu de cette classe". As soon as set theory has been formalized as a mathematically theory e.g. based on axioms of some formal language, the question is whether in the universe of sets given by the axioms there are always choice-sets or not. We believe that the ZF-axioms describe in a correct way our intuitive contemplations concerning the notion of set. The axiom of choice (AC) is intuitively not so clear as the other ZF-axioms are, but we have learned to use it because it seems to be indispensable in proving mathematical theorems. On the other hand the (AC) has "strange" consequences, such as "every set can be well-ordered" (Zermelo, Math. Ann. 59(1904) and Math. Ann. 65(1908)) and we are unable to "imagine" a well-ordering of the set of real numbers (is this only a question of concentration when thinking?) Hence there are two important questions:

(1) Is the axiom of choice (AC) provable of refutable in ZF?
Or (2) is (AC) at all events consistent with ZF?

The same questions may be asked in the case of the Continuum Hypothesis: (CH): $\neg\bigvee_x(\aleph_0 < \overline{\overline{x}} < 2^{\aleph_0})$, $\overline{\overline{x}}$ is the cardinality of x, hence (assuming (AC)): (CH) $\leftrightarrow 2^{\aleph_0} = \aleph_1$. The hypothesis (CH) was the simplest conjecture concerning the cardinality of the continuum of G.Cantor who spent many years in order to prove it without obtaining a proof.

The generalization of (CH) is:

(GCH) General Continuum Hypothesis: $\bigwedge_x \neg \bigvee_y(\overline{\overline{x}} < \overline{\overline{y}} < \overline{\overline{2^x}})$;

(AH) Aleph - Hypothesis: $\bigwedge_\alpha(2^{\aleph_\alpha} = \aleph_{\alpha+1})$.

These assumptions are often used (mostly in Topology and Algebra) in order to decide questions about the cardinality of certain spaces or structures. As in the case of (AC), the (GCH) has unexpected consequences. We give an example (due to W.Sierpiński (Rendiconti del Circolo Mat. di Palermo, Serie II, vol.1(1952)p.6-10)):

"The euklidean space E of 3 dimensions is a union of sets E_1, E_2, E_3 such that whenever A is a line parallel to the ith axis of coordination (i = 1,2,3), then $A \cap E_i$ is finite"

Classically it is known, that (GCH) implies (AC) and (AH), see the important survey (in particular p.314):

[53] A.LINDENBAUM-A.TARSKI: Communication sur les recherches de la Théorie des ensembles; Comptes Rend. de Séances de la Soc. Sci. et lettres, Varsovie 19(1926), Classe III, p.299-330.

This paper does not contain any proofs. The first proof for (GCH) $\to$ (AC) (in ZF^0) was published by W.Sierpinski (Fund. Math. 33(1945) p.137-168). Another proof has been given by E.P.Specker (Archiv der Math. 5(1954) p.332-33 . In full ZF (id est ZF^0 + regularity), (GCH) and (AH) are equivalent, since (AH) obviously implies that every set of type $P(\aleph_\alpha)$ can be well-ordered [$x \leqslant y \leftrightarrow Min((x \cup y) - (x \cap y)) \in y$ is a total ordering of $2^{\aleph_\alpha}$] , which in turn implies the axiom of choice as H.Rubin has shown in 1960 (see the Book "Equivalents of the (AC)" by H. and J.E.Rubin, Amsterdam 1963, page 77).

In 1936/37 Gödel in his lectures given at the University of Vienna (Austria) has proved the consistency of (AC) with ZF. Gödel has extended this result by showing that the (GCH) is consistent with ZF too. The first announcement of this result appeared in 1938:

[23] K.GÖDEL: The Consistency of the Axiom of Choice and the Generalized Continuum Hypothesis; Proc. Nat. Acad. Sc.24(1938)p. 556-557.

A more detailed version appeared in 1939:

[24] K.GÖDEL: Consistency-Proof for the Generalized Continuum-Hypothesis; Proc. Nat. Acad. Sci. USA, 25(1939)p.220-225.

The proof in [24] consists in defining by means of a transfinite Hiearachy M_α a semantical model of ZF + (AC) + (GCH). In lectures presented at Princeton 1938/39 Gödel has proved the Consistency by means of a syntactic model of NBG + (AC) + (GCH) in NBG. The syntactic model Δ is defined by means of a hierarchy F_α of constructible sets. Gödel's Princeton lecture notes appeared first in 1940 in print:

[25] K.GÖDEL: The consistency of the axiom of choice and of the generalized Continuum-Hypothesis; Princeton University Press 1966 (7th printing), Annals of Mathematics Studies Nr. 3.

In our presentation of the consistency result we shall follow [24] but with some modifications due to R.B.Jensen (unpublished Seminar notes, Bonn 1967) and to C.Karp:

[42] C.KARP: A proof of the relative consistency of the Continuum Hypothesis; in: "Sets, Models and Recursion Theory", Proceedings Leicester 1965, North-Holland Publ. Comp. Amsterdam 1967.

B) THE CONSTRUCTION OF THE MODEL

Definition. A model $\mathfrak{M} \doteq \langle M,E \rangle$ of set theory is called "standard" iff M is transitiv (in the sense of the meta-theory) and the binary relation E coincides with the actual membership relation $\in$ when restricted to M.

The model L will be constructed by means of a hierarchy of sets M_α. The construction may be motivated as follows:

Let A be a set of nonempty, pairwise disjoint sets. A choice set C for A is a subset of $\bigcup A$, hence an element of $P(\bigcup A)$, the power set of $\bigcup A$. Hence one might think, that the axiom of choice is provable from the axioms of union and Powerset. But the problem is, in order to exhibit such a choice set we need to have a ZF-formula $\Phi(x)$ which, by the Aussonderungsaxiom, describes one choice set. We do not know in general, how to obtain such a description. In order to get that (AC) holds in our model, we shall construct it by taking only those sets which are definable in a certain ramified language.

Definition: Let $\mathcal{L}$ be a first order language (allowing all manner of relations, functions, predicates and individual constants to occur in the alphabet of $\mathcal{L}$) and let $\mathfrak{A} \triangleq \langle A, \ldots \rangle$ be a relational system appropriate for $\mathcal{L}$, with domain A. A subset S of A is said to be definable in $\mathfrak{A}$ iff there is a formula Φ of $\mathcal{L}$ with one free variable x, such that $S \triangleq \{a;\ a \in A\ \&\ \mathfrak{A} \models \Phi[a]\}$. Let Def($\mathfrak{A}$) be the set of subsets definable in $\mathfrak{A}$.

Theorem (K.Gödel): Every standard model $\mathfrak{M}$ of ZF contains a submodel $\mathfrak{N}$ with the same ordinals in which ZF + (AC) + (GCH) hold.

Proof: Define in $\mathfrak{M}$ a hierarchy M_α of sets for ordinals α of $\mathfrak{M}$ in the following way:

$$M_0 = \emptyset$$
$$M_\lambda = \bigcup \{M_\alpha;\ \alpha < \lambda\} \text{ if } \lambda \text{ is a limit ordinal,}$$
$$M_{\alpha+1} = \mathrm{Def}(M_\alpha)$$

where Def(M_α) denotes the set of all parametrically definable elements and subsets of the structure $\langle M_\alpha, \in, \emptyset \rangle$, or equivalently the set of all elements of M_α and subsets of M_α which are definable in the structure $\langle M_\alpha, \in, x \rangle_{x \in M_\alpha} = \mathfrak{M}_\alpha$.
The language $\mathcal{L}_\alpha$ of $\mathfrak{M}_\alpha$ has besides the ordinary ZF-symbols individual constants $\underline{x}$ for each $x \in M_\alpha$ and $\underline{x}$ is interpreted in $\mathfrak{M}_\alpha$ by x.
Define finally:

$$L = \bigcup \{M_\alpha;\ \alpha \in On^{\mathfrak{M}}\}$$

where $On^{\mathfrak{M}}$ is the class of all ordinals in $\mathfrak{M}$. A set is called constructible iff it is a member of some M_α. Since the M_α's are defined in $\mathfrak{M}$ they all are subsets of $\mathfrak{M}$ and hence $L \subseteq \mathfrak{M}$. L is called the $\mathfrak{M}$-class of constructible sets of $\mathfrak{M}$. We claim that in L the axioms of ZF and (AC) and (GCH) hold. This will be proved in several steps. First notice, that $\alpha \leq \beta$ implies $M_\alpha \subseteq M_\beta$. Furthermore: if $\alpha < \beta$ then $M_\alpha \in M_\beta$. This holds since obviously $M_\alpha \in M_{\alpha+1} \subseteq M_\beta$. For a constructible set let ord(x) be the least ordinal α such that $x \in M_\alpha$. Obviously ord(x) can never be a limit ordinal.

Lemma: If x is constructible and $y \in x$, then y is constructible and ord(y) < ord(x). The sets M_α are thus transitive.

Lemma: Every ordinal α of $\mathfrak{M}$ is constructible and ord(α) = α + 1.

Proof: First we show that $\mathrm{ord}(\alpha) \nleq \alpha$. Suppose there are constructible ordinals α such that $\mathrm{ord}(\alpha) = \beta \leq \alpha$. Then $\alpha \in M_\beta \subseteq M_\alpha$, hence there are constr. ordinals α such that $\alpha \in M_\alpha$. Let α_0 be the least such ordinal. If $\alpha_0 = \gamma + 1$, then $\alpha_0 \in M_{\alpha_0} = \mathrm{Def}(M_\gamma)$, hence either $\alpha_0 \in M_\gamma$ or $\alpha_0 \subseteq M_\gamma$. But since $\gamma \in \alpha_0$ and M_γ is transitiv, we obtain (in both cases): $\gamma \in M_\gamma$ - a contradiction, α_0 would not be minimal. If $\alpha_0 = \bigcup \alpha_0$, then $\alpha_0 \in M_\rho$ for some $\rho < \alpha_0$, id est $\rho \in \alpha_0$. Then $\rho \in M_\rho$ since M_ρ is transitiv; again a contradiction. Hence we always must have $\mathrm{ord}(\alpha) \geq \alpha + 1$. We shall show that $\mathrm{ord}(\alpha) = \alpha + 1$. Let α_0 be the first ordinal (if such ordinals exist) such that $\alpha_0 \notin M_{\alpha_0+1}$. For every ordinal $\beta \in \alpha_0$ therefore $\beta \in M_{\beta+1}$, which is $\beta \in \mathrm{Def}(M_\beta)$. Hence, if α_0 is a limit ordinal: $\alpha_0 \subseteq \bigcup\{M_\beta;\ \beta \in \alpha_0\} = M_{\alpha_0}$. In order to show that $\alpha_0 \in M_{\alpha_0+1}$ holds, we have to show that α_0 is a definable subset of M_{α_0}. Consider the formula $\Phi(x)$:

$$\bigwedge_u \bigwedge_v [u \,\varepsilon\, x \wedge v \,\varepsilon\, x \rightarrow u \,\varepsilon\, v \vee u = v \vee v \,\varepsilon\, u] \wedge \bigwedge_u \bigwedge_v [v \,\varepsilon\, u \wedge u \,\varepsilon\, x \rightarrow v \,\varepsilon\, x]$$

(x is totally ordered by $\in$ and transitiv). $\Phi(x)$ defines a certain subset A of M_{α_0}. Since M_{α_0} is transitive and ε is interpreted by the actual membership relation $\in$, A must be precisely the set of those ordinals, which are in M_{α_0}, But these are by definition at least the elements β of α_0. If $\gamma \in M_{\alpha_0}$ for $\gamma \geq \alpha_0$, then by the transitivity of M_{α_0}, $\alpha_0 \in M_{\alpha_0}$ which cannot happen as it was shown above. Hence A consists precisely of the elements of α_0. Id est $\alpha_0 \hat{=} A$ and α_0 is definable in M_{α_0} and hence an element of M_{α_0+1}.

If α_0 is a successor ordinal, say $\alpha_0 = \beta + 1$, then $\alpha_0 \hat{=} \beta \cup \{\beta\}$. Since $\beta \in M_{\beta+1} = M_{\alpha_0}$ (by minimality of α_0), hence definable, we have obviously $\alpha_0 \in M_{\alpha_0+1}$ too, q.e.d.

Since $M_0 = \emptyset$, hence $\emptyset \in M_1$, the axiom of Null set is true in L. Since extensionality holds in $\mathfrak{M}$ and elements of constructible elements are constructible too, the axiom of extensionality holds in L. The axiom of pairs is also obviously true in L.

Lemma: The axiom of sum-set holds in L.

Proof: Let a be a constructible set and α an ordinal of $\mathfrak{M}$ such that $a \in M_\alpha$. Since M_α is transitive $b = \bigcup\{x;\ x \in a\}$ is contained (as a subset) in M_α. The formula $\bigvee_y (y \,\varepsilon\, \underline{a} \wedge x \,\varepsilon\, y)$ with constant $\underline{a}$ from M_α defines b as a subset of M_α. Hence $b \in \mathrm{Def}(M_\alpha) = M_{\alpha+1} \subseteq L$.

Lemma: The axioms of regularity and infinity hold in L.

Proof. We have shown, that every ordinal of $\mathfrak{M}$ is constructible. Since the axiom of Infinity holds in $\mathfrak{M}$, ω is in $\mathfrak{M}$, hence in L. But ω satisfies the axiom of Infinity in L. Since the axiom of regularity holds in $\mathfrak{M}$ and $L \subseteq \mathfrak{M}$ and the epsilon-relations of L and $\mathfrak{M}$ coincide, the axiom of regularity must hold in L as well.

Lemma: The power set axiom holds in L.

Proof. Let a be a set of L and b be the set of constructible subsets of a. Since a is a set of $\mathfrak{M}$, b is a subset of the powerset of a in the sense of $\mathfrak{M}$. Therefore b is a set of $\mathfrak{M}$ (since the hierarchy M_0, M_1,... M_α,... was defined in $\mathfrak{M}$). Define in $\mathfrak{M}$ the mapping $x \to ord(x)$ from elements of b to ordinals. Since the axiom of replacement is true in $\mathfrak{M}$, $\{ord(x);\ x \in b\}$ has a least upper bound α. Hence $b \subseteq M_\alpha$. Let $\beta = Max\{\alpha, ord(a)\}$, then $b \subseteq M_\beta$ and $a \in M_\alpha$. The formula $\bigwedge_u (u \in x \to u \in \underline{a})$ with one free variable x and one constant $\underline{a}$ from M_β defines b as a subset of M_β. Thus $b \in M_{\beta+1}$ and b is constructible./ The proof of the following lemma uses techniques which are essentially due to E.J.Thiele in his 1961 paper, rediscovered in 1963 by P.J.Cohen.

Lemma: The axiom of replacement holds in L.

Proof. Consider a formula $\Phi(x,y,\underline{a}_1,\dots,\underline{a}_k)$ with free variables x and y and parameters $a_1,\dots,a_k$ which are in L. Suppose that for every x in L there is precisely one y in L such that $\Phi(x,y,\underline{a}_1,\dots\underline{a}_k)$. Now let a be any set in L and let b be the collection of constructible sets y such that there are sets x with $\Phi(x,y,\dots)$ and $x \in a$. We have to show that b is constructible. Every element y of b is constructible and since b is a set in the sense of $\mathfrak{M}$, $\{ord(y);\ y \in b\}$ has by the axiom of replacement in $\mathfrak{M}$ a supremum α. Define $\beta_0 = Max\{\alpha, ord(a), ord(a_1),\dots, ord(a_k)\}$. Then $a, a_1,\dots,a_k \in M_{\beta_0}$ and $b \subseteq M_{\beta_0}$. We shall show that $b \in M_{\gamma_0+1}$. This can be done in the most simplest way by applying (in $\mathfrak{M}$) the following generalized refection principle:

(GRP): Let $\{W_\alpha;\ \alpha$ an ordinal$\}$ be a nested sequence of sets W_α such that for $\lambda = \bigcup\lambda$ it holds that $W_\lambda = \bigcup\{W_\alpha;\ \alpha < \lambda\}$. Let $W = \bigcup\{W_\alpha;\ \alpha$ an ordinal$\}$. If $\Phi(x_1,\dots,x_n)$ is a formula of ZF, then for every ordinal α there is a $\beta > \alpha$ such that

$$\bigwedge_{x_1}\dots\bigwedge_{x_n}[x_1,\dots,x_n \in W_\beta \to (Rel(W,\Phi) \leftrightarrow Rel(W_\beta,\Phi))].$$

It can be shown exactly in the same way as we have proved the principle of complete reflection (CR'), that every instance of (GPR) is a theorem of ZF (from the hierarchy V_α we needed only the facts, that for λ a limit ordinal: $V_\lambda = \bigcup \{V_\alpha ; \alpha < \lambda\}$ and $\alpha \leq \beta \to V_\alpha \subseteq V_\beta$). Since $\mathfrak{M}$ is by assumption a model of ZF, (GPR) holds in $\mathfrak{M}$. We shall apply (GPR) for the hierarchy M_α and the formula $\Phi(x,y,x_1,\ldots,x_k)$ with free variables $x_1,\ldots,x_k$ replaced for the constants $a_1,\ldots,a_k$ (alphabetic change of bound variables in Φ may be needed). Hence for the ordinal β_0 previously defined and Φ there is and ordinal $\gamma_0 > \beta_0$ such that

$$\bigwedge_x \bigwedge_y \bigwedge_{x_1} \ldots \bigwedge_{x_k} [x,y,x_1,\ldots,x_k \in M_{\gamma_0} \to (\mathrm{Rel}(L,\Phi) \leftrightarrow \mathrm{Rel}(M_{\gamma_0},\Phi))].$$

Since $\mathrm{ord}(a_1) \leq \beta_0 < \gamma_0, \ldots, \mathrm{ord}(a_k) \leq \beta_0 < \gamma_0$ we have that $a_1,\ldots,a_k$ are elements of M_{γ_0}; hence we obtain:

$$\bigwedge_x \bigwedge_y [x,y \in M_{\gamma_0} \to (\mathrm{Rel}(L,\Phi(x,y,\underline{a}_1,\ldots,\underline{a}_k)) \leftrightarrow \mathrm{Rel}(M_{\gamma_0},\Phi(x,y,a_1,\ldots,\underline{a}_k)))].$$

For $\Phi(x,y,\underline{a}_1,\ldots,\underline{a}_k)$ let us simply write $\Phi(x,y)$ in the sequel. The interpretation of $\Phi(x,y)$ in M_{γ_0} consists of those pairs $\langle x,y\rangle \in M_{\gamma_0} \times M_{\gamma_0}$ such that $\mathrm{Rel}(M_{\gamma_0},\Phi(x,y))$. By the last formula it is equivalent to say that the interpretation of $\Phi(x,y)$ in M_{γ_0} consists of those pairs $\langle x,y\rangle \in M^2_{\gamma_0}$ such that $\mathrm{Rel}(L,\Phi(x,y))$. Hence the formula

$$\bigvee_x (x \in \underline{a} \wedge \Phi(x,y))$$

with one free variable y and constants $\underline{a},\underline{a}, \ldots,\underline{a}_k$ defines the set b in M_{γ_0}. Thus $b \in M_{\gamma_0+1}$ and b is constructible, q.e.d.

Remark. The proof given above that $\langle L,\in\rangle$ is a model of ZF is carried out informally. A detailed proof defines first in $\mathfrak{M}$ a language $\mathcal{L}_{\mathfrak{M}}$ which contains besides the usual ZF-symbols a constant $\underline{x}$ for each set x of $\mathfrak{M}$. In order that $\mathcal{L}_{\mathfrak{M}}$ is a class of $\mathfrak{M}$ one has to arrange that all the symbols of the alphabet of $\mathcal{L}_{\mathfrak{M}}$ are sets of $\mathfrak{M}$. R.B.Jensen in his lecture notes (Springer, 1967) "Modelle der Mengenlehre" gives a detailed solution. The satisfaction relation $\models$ between structures $\langle s,\in \restriction s\rangle$ (which are sets of $\mathfrak{M}$) and formulae of $\mathcal{L}_{\mathfrak{M}}$ can be defined in $\mathfrak{M}$ (see Mostowski's "Constructible sets"). Then the class L can be defined in $\mathfrak{M}$ by means of transfinite recursion.

In order to show, that $\langle L,\in \rangle$ is a model of (AC) and (GCH) we shall show that a much stronger axiom, the socalled axiom of constructibility" V = L", holds in $\langle L,\in \rangle$ and that "V = L" implies (AC) as well as (GCH). So far we have only shown, that "L" can be defined within any given standard model $\mathfrak{M}$. Now we shall show that the statement informally expressed as "every set is constructible:" V = L" can be expressed by a formula of the (pure) ZF-language, so that V = L is a certain ZF-sentence. We shall prove a stronger fact, namely that V = L is in ZF equivalent to a Σ_1-formula (this result is due to C. Karp [42]).

c) Σ_1^{ZF}-FORMULAE

Defintion. (A.Lévy): A formula Φ of the ZF-language is a Σ_0-formula iff Φ contains no unbounded quantifiers; Φ is Σ_1, iff Φ has the form $\bigvee_x \Psi$ with Ψ a Σ_0-formula; Φ is Π_1 iff Φ has the form $\bigwedge_x \Psi$ for a Σ_0-formula Ψ. If a ZF-formula Γ is in ZF (provably) equivalent to a Σ_0 (Σ_1, Π_1) formula, then Γ is called a Σ_0^{ZF} (Σ_1^{ZF}, Π_1^{ZF} respectively) formula. Γ is Δ_1^{ZF} iff Γ is both Σ_1^{ZF} and Π_1^{ZF}.

Lemma 1: If Φ and Ψ are Σ_1^{ZF}, then so are $\Phi \wedge \Psi$, $\Phi \vee \Psi$, $\bigvee_x \Phi$, $\bigwedge_x (x \in z \rightarrow \Phi)$ and $\bigvee_x (x \in z \wedge \Phi)$.

Proof. By hypothesis: $ZF \vdash \Phi \leftrightarrow \bigvee_y \Psi$ with $\Psi \Sigma_0$. We have to show that $ZF \vdash \bigvee_x \bigvee_y \Psi \leftrightarrow \bigvee_z (\bigvee_{x\in z} \bigvee_{y \in z} \Psi)$. The part "$\leftarrow$" is obvious; for the part "$\rightarrow$" take $z = \{x,y\}$, z a Variable not in Ψ. Hence $\bigvee_x \Phi$ is Σ_1^{ZF}. Now it follows that also $\Phi \wedge \Psi$ and $\Phi \vee \Psi$ are Σ_1^{ZF}. Next we show that:

(*) $ZF \vdash \bigwedge_x (x \in z \rightarrow \bigvee_y \Psi) \leftrightarrow \bigvee_u \bigwedge_{x\in z} \bigvee_{y\in u} \Psi$.

The part "$\leftarrow$" is obvious. Ad "$\rightarrow$": assume x,y are free in Ψ. Define a function F:

$F(x) = \{y;\ \Psi(x,y,\ldots) \wedge \rho(y) = \mathrm{Min}\{\alpha;\ \bigvee_v (\rho(v) = \alpha \wedge \Psi(x,v,\ldots))\}\}$

By the axioms of sumset and replacement (*) follows with $u = \bigcup\{F(x);\ x \in z\}$. It remains to show that $\bigvee_x (x \in z \wedge \Phi)$ is Σ_1^{ZF} for $\Phi \triangleq \bigvee_y \Psi$ with $\Psi \Sigma_0$. But $x \in z \wedge \Psi$ is Σ_0, hence $\bigvee_y (x \in z \wedge \Psi)$ is Σ_1 and therefore $\bigvee_x (\bigvee_y x \in z \wedge \Psi)$ is Σ_1^{ZF} as was shown above.

Corallary 2: If Φ and Ψ are Δ_1^{ZF}, so are $\Phi \wedge \Psi$, $\Phi \vee \Psi$, $\Phi \rightarrow \Psi$, $\bigwedge_x (x \in z \rightarrow \Phi)$ and $\bigvee_x (x \in z \wedge \Phi)$.

Lemma 3: If $ZF \vdash \bigvee_z \bigvee_n \bigwedge_x (\Phi(x) \rightarrow x \in \bigcup^n z)$, then

(a) $\bigwedge_x (\Phi \rightarrow \Psi) \leftrightarrow \bigwedge_{y_1 \in z} .. \bigwedge_{y_n \in y_{n-1}} \bigwedge_{x \in y_n} (\Phi(x) \rightarrow \Psi)$

(b) $\bigvee_x (\Phi \rightarrow \Psi) \leftrightarrow \bigvee_{y_1 \in z} \dots \bigvee_{y_n \in y_{n-1}} \bigvee_{x \in y_n} (\Phi \rightarrow \Psi)$,

where it is supposed that $y_1, \dots y_n$ are not free in Φ, Ψ.

Proof. "$\rightarrow$" obvious. "$\leftarrow$": since $x \in \bigcup^n z \leftrightarrow \bigvee_{y_1 \in z} \dots \bigvee_{y_n \in y_{n-1}} (x \in y_n)$ the lemma follows immediately.

Lemma 4: The following formulae are Σ_0^{ZF}-formulae: (1): $x = y$, (2): $y = \{x,z\}$, (3): $y = \langle x_1, \dots, x_n \rangle$, (4): $y = x \cap z$, $y = \bigcap x$, (5): $y = x \cup y$, $y = \bigcup x$, (6): $y = x - z$, (7): $y = Dom(f)$, $y \in Dom(f)$, (8): $y = Rg(f)$, $y \in Rg(f)$, (9): $y = f \upharpoonright z$, (10): $y = f''z$, (11): $x = y \times z$, (12): $y = f^{-1}$, (13): $Fnc(f)$, (14): $y = x \cup \{x\}$ and (15): $Ord(x)$ (x is an ordinal).

Here $\langle x_1, x_2 \rangle$ means the ordered pair, $-$ set theoretical difference, $Dom(f)$ the domain of f, $Rg(f)$ the range of f, $f \upharpoonright z$ the restriction of f to z, $f''z = \{f(y); y \in z\}$ the image of z under f for $z \subseteq Dom(f)$, $\times$ cartesian product and $Fnc(f)$ expresses that f is a function.
(1) and (2) are immediate, (3) follows from (2), (4) follows from lemma 3; (5) and (6) are obvious, (7) uses (3) and lemma 3, (8) is analoguous to (7); (9), (10), (12) and (13) are proved by means of lemma 3. In all cases use appropriate defining formulas (see C.Karp [42] e.g.).

Lemma 5: If $\Phi(y, x_1, \dots, x_n)$ is Σ_1^{ZF} and $ZF \vdash \bigwedge_{x_1} \dots \bigwedge_{x_n} \overset{1}{\bigvee_y} \Phi(y, x_1, \dots, x_n)$. then Φ is Δ_1^{ZF}.

Proof. Since $ZF \vdash \neg \Phi \leftrightarrow \bigvee_z (z \neq y \wedge \Phi(z, ..))$, $\neg \Phi$ is Σ_1^{ZF}, hence Φ is Π_1^{ZF}.

Lemma 6: Substitutionsprinciple: Let $G_1, \dots, G_n$ be m-place functions such that $y_i = G_i(x_1, \dots, x_m)$ is Σ_1^{ZF}. Then if $\langle u_1, \dots, u_n \rangle \in R$ is Δ_1^{ZF},

$$\langle G_1(x_1, \dots, x_m), \dots, G_n(x_1, \dots, x_m) \rangle \in R$$

is Δ_1^{ZF} too.

Proof. By lemma 1 the proof of the claim follows from
$\langle G_1(\vec{x}),\dots,G_n(\vec{x})\rangle \varepsilon R \leftrightarrow \bigvee_{y}\dots\bigvee_{y_n} (\langle y_1,\dots,y_n\rangle \varepsilon R \wedge \bigwedge_{i\leq n} y_i = G_i(\vec{x}))$
and the formula obtained by replacing (in both places of occurence) ε by $\not\varepsilon$.

Lemma 7: Let $x = G(y_0,\dots,y_n)$ and $z = H(y_1,\dots,y_n)$ be n + 1 resp. n - place Σ_1^{ZF}-functions and $\langle y_0,\dots,y_n\rangle \varepsilon R$ be a n + 1 - place Δ_1^{ZF}-relation, then F is a n-place Σ_1^{ZF}-function, where $F(y_1,\dots,y_n) = \{G(y_0,\dots,y_n);\ y_0 \varepsilon H(y_1,\dots,y_n) \wedge \langle y_0,\dots,y_n\rangle \varepsilon R\}$.

Lemma 8: Ordinal recursion: Let $x = G(y_0,\dots,y_n)$ be a n + 1-place Σ_1^{ZF}-function and F the function defined by ordinal recursion on y_1 as $F(y_1,\dots,y_n) = o$ if $\neg\, Ord(y_1)$, $F(y_1,\dots,y_n) = G(F\restriction y_1, y_1,\dots,y_n)$ if $Ord(y_1)$, then F is Σ_1^{ZF}.

For a proof of Lemmata 7, 8 the reader is referred to Jensen or Karp [42] p.25-26. $F\restriction y_1$ = the function whose domain is y_1 and whose values are $F(\xi, y_2,\dots,y_n)$ for $\xi \varepsilon y_1$.

Notice that the formulae Ord(x), Lim(x), $x \varepsilon \omega$, $x = \omega$ are Σ_0^{ZF}. By ordinal recursion it follows that $z = x + y$ (ordinal addition) is Δ_1^{ZF}, and analogously $z = x.y$ and $z = x^y$ (ordinal multiplication, exponentiation). Further $y = C(x)$, where $C(x) = x \cup \{C(z);\ z \varepsilon x\}$ is the transitive closure of x, is Δ_1^{ZF} (by lemma 4, lemma 6 and lemma 8).

D) Σ_1^{ZF}-definability of syntactical notions

A sequence of symbols is defined to be a finite sequence (function) such that $Z = \{y;\ Fnc(y) \wedge Dom(y)\varepsilon\omega\}$ is the class of all sequences of symbols. $z \varepsilon Z$ is Δ_1^{ZF} (by lemma 4,6,8). Lemma 3 and since $x \varepsilon \omega$ is Σ_0^{ZF} gives moreover that $z \varepsilon Z$ is Σ_0^{ZF}. If $Z_u = \{y \varepsilon Z;\ Rg(y) \subseteq u\}$, then $y \varepsilon Z_u$ and $y = Z_u$ are both Δ_1^{ZF}. Let uv denote concatenation of the sequences u,v; then $z = uv$ is Δ_1^{ZF}.

Arithmetization

In order to enable us to consider the formulas of the ZF-object language as finite sequences we must identify the Alphabet of the object language with special finite sequences:

$x_i = \{\langle\langle i,1\rangle , 0\rangle\}$ (x_i is the ith variable),

$\dot{\neg} = \{\langle\langle 0,2\rangle , 0\rangle\}$

$\dot{\wedge} = \{\langle\langle 1,2\rangle , 0\rangle\}$ $\dot{=} = \{\langle\langle 5,2\rangle , 0\rangle\}$

$\dot{\bigwedge} = \{\langle\langle 2,2\rangle , 0\rangle\}$ $(= \{\langle\langle 6,2\rangle , 0\rangle\}$

$E^{\cdot} = \{\langle\langle 3,2\rangle , 0\rangle\}$ $) = \{\langle\langle 7,2\rangle , 0\rangle\}$

$\dot{\varepsilon} = \{\langle\langle 4,2\rangle , 0\rangle\}$ $\underline{x} = \{\langle\langle x,0\rangle , 0\rangle\}$

($E^{\cdot}$ is a comprehension operator: ensemble des...)

For every set $x \in V$ we have hence a constant $\underline{x}$ in our object-language The symbols of the Alphabet are by definition (above) functions with domain $\{0\}$ and values $\langle i,g\rangle$, $i,g \in \omega$. The formulae $y = x_i$, $y \in Vbl$ (y is a variable), $y = \underline{x}$, $y \in Konst$ (y is a constant of form $\underline{x}$) are all Σ_0^{ZF}. The class Fml of all formulae is defined by recursion (as usual) and we get that $x \in Fml$ is Δ_1^{ZF} (using lemma 8). The replacement of a variable z by a constant t in a formula $\phi^{\cdot}$, in symbols $\phi^{\cdot}(^z/_t)$, is Δ_1^{ZF}-definable. If $Fr(\phi^{\cdot})$ is the set of free variables in $\phi^{\cdot}$, then $y = Fr(\phi^{\cdot})$ is Δ_1^{ZF}, hence $y \in S$ (y is a sentence, i.e. has no free variables) is Δ_1^{ZF} too. Finally let Fml_u be the set of formulae whose subformulas are elements of u. Then $y \in Fml_u$ and $y = Fml_u$ are Δ_1^{ZF}. A similar result holds for sets S_u of sentences whose subformulae are in u.

E) Δ_1^{ZF}-definability of semantic notions

For any set u and any sentence $\phi^{\cdot} \in S_u$ the satisfaction relation $u \models \phi^{\cdot}$ is defined as usual by recursion: $u \models \underline{x}\dot{\varepsilon}\underline{z} \leftrightarrow x \varepsilon z$, $u \models \underline{x} \dot{=} \underline{z} \leftrightarrow x = z$, $u \models \phi^{\cdot} \dot{\wedge} \psi^{\cdot} \leftrightarrow u \models \phi^{\cdot} \wedge u \models \psi^{\cdot}$, $u \models \dot{\neg}\phi^{\cdot} \leftrightarrow \neg\, u \models \phi^{\cdot}$, $u \models \dot{\bigwedge}_x \phi^{\cdot} \leftrightarrow \bigwedge_{x \varepsilon u} u \models \phi^{\cdot}(^x/_{\underline{x}})$ and finally

$u \models \underline{x} \varepsilon E^{\cdot} z \phi^{\cdot} \leftrightarrow u \models \phi^{\cdot}(^z/_{\underline{x}})$.

As a consequence we get that $u \models \phi^{\cdot}$ is Δ_1^{ZF} (this is shown by proving that that the relation: $\phi^{\cdot}$ is a u-component (= subformula) of $\psi^{\cdot}$ is well-founded, and Δ_1^{ZF} and then applying lemma 7).

Define KT_u to be the set of all class-terms in u:

$KT_u = \{y;\ \bigvee_{\phi^{\cdot}} \phi^{\cdot} \varepsilon Fml_u \wedge \bigvee_z (z \varepsilon Vbl \wedge y = E^{\cdot} z \phi^{\cdot} \wedge Fr(\phi^{\cdot}) = \{z\})\}$.

Then $y \varepsilon KT_u$ and $y = KT_u$ are both Δ_1^{ZF}.
Now let $u[t]$ be the set $\{x;\ x\varepsilon u \wedge u \models \phi'(^z/_{\underline{x}})\}$ for the comprehension term $t = E'z\phi'\varepsilon KT_u$. Then $y = u[t]$ is Δ_1^{ZF} (the u-valuation of t)

Lemma 9: $y = Def(u)$ is Δ_1^{ZF}.

Proof. $Def(u)$ is defined to be the set $\{u[t]\ ;\ t\varepsilon KT_u\}$. Since $a = KT_u$ is Δ_1^{ZF} the lemma follows from the fact that $y = u[t]$ is Δ_1^{ZF} using lemma 7.

F) The Σ_1^{ZF}-definability of the constructible model

The class L (i.e. the constructible, inner model) is defined by the following recursion:

$$L(x) = \begin{cases} \emptyset & \text{if } x = \emptyset, \\ Def(L(z)) & \text{if } Ord(x)\wedge Ord(z)\wedge x = z+1, \\ \bigcup\{L(z);\ z\varepsilon x\} & \text{if } Lim(x) \\ \emptyset & \text{otherwise} \end{cases}$$

and $L = \bigcup\{L(x);\ Ord(x)\}$.

Lemma 10. $y = L(x)$ is Δ_1^{ZF}.

Proof. Define the following 2-place function $G(x,u)$ by: $G(x,u) = \emptyset$ if $x = \emptyset$, $G(x,u) = Def(u(z))$ if $Ord(z)\wedge x = z+1\wedge Fnc(u)\wedge z\varepsilon Dom(u)$, $G(x,u) = \bigcup\{u(z);\ z\varepsilon x\}$ if $Lim(x)\wedge Fnc(u)\wedge x \subseteq Dom(u)$, $G(x,u) = \emptyset$ otherwise.

Obviously: $L(x) = G(x,L\restriction x)$ and Lemma 4 yields $v = u(z)$ is Δ_1^{ZF}, hence by lemma 6: $y = Def(u(z))$ is Δ_1^{ZF}, since $y = Def(v)$ is Δ_1^{ZF} by lemma 9. Thus lemma 7 gives us that $y = G(x,u)$ is Δ_1^{ZF} and lemma 8 yields that $y = G(x,L\restriction x)$ is Δ_1^{ZF} too.

Corollary 11: $z\varepsilon L(x)$ is Δ_1^{ZF}.

Proof: Since $z\varepsilon y$ is Δ_1^{ZF}, the corollary follows from lemma 6 and 10.

Corollary 12: $z\varepsilon L$ is Σ_1^{ZF}.

Proof: Since $z\varepsilon L \leftrightarrow \bigvee_x (Ord(x)\wedge z\varepsilon L(x))$ the corollary follows from lemma 1, lemma 4 and corollary 11.

G) Properties of the class L of constructible sets

Theorem (Gödel [25]): There is a ZF-formula $\Phi(x,y)$ which wellorders the class L.

Outline of proof. Remark that by induction one can show that each L(x), for Ord(x), has a definable wellordering W(x). Obviously $\emptyset = W(0)$ is such a wellordering of L(0). Suppose we have defined W(x) for all ordinals $x < \alpha$. If Lim(α), then
$W(\alpha) = \{\langle x,y\rangle;\ \bigvee_\beta \bigvee_\gamma [(\beta < \gamma \wedge x \varepsilon L(\beta) \wedge y \varepsilon L(\gamma) \wedge \neg\, y \varepsilon L(\beta)) \vee$

$\vee (\beta = \gamma \wedge \langle x,y\rangle \varepsilon W(\beta))\}$ wellorders $L(\alpha)$. If $\alpha = \beta+1$, then

$L(\alpha) = Def(L(\beta))$; the elements $L(\alpha)[t]$ of $L(\alpha)$ are definable subsets (or elements) of $L(\beta)$. The language $Fml_{L(\beta)}$ which describes these subsets has besides the wellordered sequence of the usual ZF-language constants $\underline{x}$ for $x \varepsilon L(\beta)$. But $W(\beta)$ induces obviously a wellordering between these constants and hence $KT_{L(\beta)}$ is wellordered. Thus the whole language $Fml_{L(\beta)}$ which is used to obtain "definable" subsets of $L(\beta)$ is wellorderable and the wellordering between the formulas can be used to obtain a wellordering of $L(\alpha) = Def(L(\beta))$. The relation $W(\alpha)$ can be defined thus in terms of $W(\beta)$ and the valuation (interpretation)-function for terms t of $TK_{L(\beta)}$ (for a rigurous proof see C. Karp [42] p. 16-17).

Corollary: The statement V = L implies the axiom of choice (AC).

Remark: V = L is an abbreviation for $\bigwedge_x x \varepsilon L$: every set is constructible.

Let $\approx$ be the relation of cardinal equivalence (equipollence). Using (AC) one shows that for infinite ordinals α
$\alpha \approx \alpha+\alpha \approx \alpha \cdot \alpha \approx \alpha^n$ (all $n \varepsilon \omega$). Therefore one is able to show by induction that $\alpha \geqslant \omega \rightarrow L(\alpha) \approx \alpha$ holds.

Lemma (Mostowski): (in ZF): If R is a binary relation on a set A such that $\langle A,R\rangle$ is well-founded and extensional, then there is a function f and a transitive set D such that f is an isomorphism of $\langle A,R\rangle$ on $\langle D,\varepsilon \upharpoonright D\rangle$. Moreover, if R is $\varepsilon \upharpoonright A$ and $C(x) \subseteq A$, then $f(x) = x$. Further, if $\Phi(x_1,\dots,x_n)$ is a Σ_0^{ZF}-formula, then

$$\bigwedge_{x_1 \varepsilon A} \dots \bigwedge_{x_n \varepsilon A} [\Phi(f(x_1),\dots,f(x_n)) \leftrightarrow Rel(A,\Phi(x_1,\dots,x_n))].$$

A relation R on A is well-founded iff $\bigwedge_B [B \neq \emptyset \to \bigvee_{x \varepsilon B} \neg \bigvee_{y \varepsilon B} \langle y,x \rangle R]$ and extensional iff $\bigwedge_{x \varepsilon A} \bigwedge_{y \varepsilon A} [\bigwedge_{z \varepsilon A} (\langle z,x \rangle \varepsilon R \leftrightarrow \langle z,y \rangle \varepsilon R) \to x = y]$.

The function f can be defined (along the well-founded and localizable (since A is a set) relation R) by recursion:

$$f(x) = \{f(y) \ ; \ \langle y,x \rangle \varepsilon R\}$$

Let D be the range of f then f is the required isomorphism. For the rest see Karp [42] p. 14. The first part of lemma was obtained by Mostowski in Fund. Math. 36 (1949) p. 143-164, see also Mostowski's book on "Constructible sets" (1969, p. 20). The additions are due to R. Montague (Bull. AMS 61 (1955), p.442) and Lévy [48].

The following theorem is contained on p. 52 (theorem 36) in:

[48] A. LÉVY: A hierarchy of formulas in set theory, Memoirs of the AMS, nr. 57 (Providence 1965).

Theorem (A. Lévy): Let Φ be a Σ_1^{ZF}-formula having no free variables except x,y. Then

$$ZF + (AC) \vdash \bigvee_x \Phi \to \bigvee_x [HC(x) \leq Max\{\aleph_0, HC(y)\} \wedge \Phi(x,y)].$$

Here HC(x) is the hereditary cardinal of x, id est: the cardinal number of the transitive closure of x: $HC(x) = \overline{\overline{C(x)}}$. In the proof one can assume that Φ is Σ_0^{ZF}, since if Φ is equivalent to $\bigvee_z \Psi(x,y,z)$ with Ψ Σ_0^{ZF}, then let $\Psi^*(u,y)$ be the following Σ_0^{ZF}-formula:

$$\bigvee_x (x \varepsilon u \wedge \bigvee_z (z \varepsilon u \wedge \Psi(x,y,z)))$$

Then take $u = \{x,z\}$ to establish $ZF \vdash \bigvee_x \Phi \leftrightarrow \bigvee_u \Psi^*$. Then the theorem for $\Psi^*(u,y)$ (with u instead of x) yields the theorem for $\Phi(x,y)$. Hence we may assume Φ to be Σ_0^{ZF}. Let $a = C(x) \cup C(y)$ and a is transitive. Since for transitive sets a and Σ_0^{ZF}-formula $\Phi(x,y)$: $\bigwedge_{x \varepsilon a} \bigwedge_{y \varepsilon a} [\Phi \leftrightarrow Rel(a,\Phi)]$ (proof by induction on the length of Φ). Hence assuming $\Phi(x,y)$ we get $Rel(a,\Phi)$ and therefore $Rel(a, \bigvee_x \Phi(x,y))$. Thus $\bigvee_x \Phi(x,y)$ is true in the structure $\langle a, \varepsilon \restriction a, y \rangle = \mathfrak{A}$. Obviously the axiom of extensionality (I) is true in $\mathfrak{A}$. By the Löwenheim-Skolem theorem here is a subset b of a such that $C(y) \subseteq b$, $\bar{\bar{b}} \leqslant \mathrm{Max}\{\aleph_0, \overline{\overline{C(y)}}\} = \mathrm{Max}\{\aleph_0, HC(y)\}$ such that $\bigvee_x \Phi(x,y) \wedge$ (I) is true in $\langle b, \varepsilon\restriction b, y\rangle$ (here (AC) is used, since the languages appropriate to these structures have besides the usual ZF-symbols a constant $\underline{s}$ for each $s\varepsilon C(y)$)

By the axiom of foundation, $\varepsilon\restriction b$ is well-founded and hence by the preceeding lemma of Mostowski there is a function f and a transitive set D such that f is an isomorphism between $\langle b, \varepsilon\restriction b\rangle$ and $\langle D, \varepsilon\restriction D\rangle$ and $f(y) = y$, id est $\langle b,\varepsilon\restriction b,y\rangle \cong \langle D,\varepsilon\restriction D,y\rangle$. Since $\bigvee_x \Phi(x,y)$ holds in $\langle b,\varepsilon\restriction b,y\rangle$, it holds also in $\langle D,\varepsilon\restriction D,y\rangle$ since Φ is Σ_0^{ZF}; hence: $Rel(D, \bigvee_x \Phi(x,y))$ and there is an $v\varepsilon D$ such that $Rel(D,\Phi(v,y))$. Since D is transitive and Φ Σ_0^{ZF} we conclude, as before, that $\Phi(v,y)$ holds. It remains to show that $HC(v) \leqslant \mathrm{Max}\{\aleph_0, HC(y)\}$. But

$$HC(v) = \overline{\overline{C(v)}} \leqslant \bar{\bar{D}} = \bar{\bar{b}} \leqslant \mathrm{Max}\{\aleph_0, HC(y)\}$$

since $v\varepsilon D$, hence $C(v) \subseteq D$ since D is transitive, further $\bar{\bar{D}} = \bar{\bar{b}}$ since f is a bijection from b onto D, q.e.d.

<u>Theorem</u> (Gödel): In ZF + (AC) it holds that for every initial ordinal $\alpha \geqslant \omega$ it holds that $P(\alpha) \cap L \subseteq L(\alpha^+)$, where $P(\alpha)$ is the (whole) powerset of α and α^+ the next cardinal after α.

<u>Proof</u>. First remark that every ordinal is constructible (same proof as in section B). If $y\varepsilon P(\alpha)$, $y \subseteq \alpha$, then $\bar{\bar{y}} \leqslant \alpha$ and $HC(y) \leqslant \alpha$. Consider the formula $Ord(x) \wedge y\varepsilon L(x)$. Since $Ord(x)$ is Σ_0^{ZF} and $y\varepsilon L(x)$ is Δ_1^{ZF} (see lemma 4 and corollary 11 in sections C and F), the formula under consideration is Σ_1^{ZF} (by lemma 1 in section C). If $y\varepsilon P(\alpha) \cap L$, then $\bigvee_x(Ord(x) \wedge y\varepsilon L(x))$ and hence by Lévy's theorem: $\bigvee_x[HC(x) \leqslant \mathrm{Max}\{\aleph_0, HC(y)\} \wedge Ord(x) \wedge y\varepsilon L(x)]$. Since $HC(y) \leqslant \alpha$ and $\omega \leqslant \alpha$, we get $HC(x) \leqslant \alpha$. But x is an ordinal, hence $\bar{\bar{x}} \leqslant \alpha$. Thus $y\varepsilon L(x)$ for $x < \alpha^+$. The sequence $L(o), L(1), \ldots, L(\beta), \ldots$ is nested, hence $L(x) \subseteq L(\alpha^+)$ and therefore $y\varepsilon L(\alpha^+)$, q.e.d.

<u>Corollary 1</u>. ZF $\vdash$ V = L $\rightarrow$ for every initial α, $P(\alpha) \subseteq L(\alpha^+)$.

<u>Corollary 2</u>. ZF $\vdash$ V = L $\rightarrow$ (GCH).

<u>Proof</u>. We know already that (AC) follows from V = L in ZF. Hence

given any set x, x is equipollent to an initial ordinal α and $P(\alpha) \subseteq L(\alpha^+)$ implies $\overline{\overline{P(x)}} = \overline{\overline{P(\alpha)}} \leq \overline{\overline{L(\alpha^+)}} = \alpha^+$. But V = L, hence $P(\alpha) \varepsilon L$ and therefore $P(\alpha) \varepsilon L(\beta)$ for some $\beta \leq \alpha^+$. But then $P(\alpha) \subseteq L(\gamma)$ for some $\gamma \leq \beta \leq \alpha^+$. But $\overline{\overline{P(\alpha)}} \leq \overline{\overline{L(\gamma)}} = \overline{\overline{\gamma}}$ and therefore $\gamma < \alpha^+$ cannot be the case (we would get $\overline{\overline{\alpha}} = \overline{\overline{x}} < \overline{\overline{P(\alpha)}} \leq \overline{\overline{\gamma}} = \overline{\overline{\alpha}}$). Hence $\gamma = \alpha^+$ and $P(x) \approx \alpha^+$ as desired, q.e.d.

So far we have shown, that (in ZF) V = L implies the (AC) and the (GCH). Now we shall return to our model constructed in section B. In order to establish that (AC) and (GCH) hold in it we need only to show, that V = L holds in it.

__Lemma__ Let $\Phi(x_1,\ldots,x_n)$ be a Σ_1^{ZF}-formula and C a transitive class, then

$$ZF \vdash \bigwedge_{x_1 \varepsilon C} \cdots \bigwedge_{x_n \varepsilon C} [\mathrm{Rel}(C,\Phi) \rightarrow \Phi].$$

Proof immediate from the fact that $\mathrm{Rel}(C,\Psi) \leftrightarrow \Psi$ for Σ_0^{ZF}-formula Ψ.

__Lemma__: If $\Phi(x_1,\ldots,x_n)$ is a Π_1^{ZF}-formula and C a transitive class, then

$$ZF \vdash \bigwedge_{x_1 \varepsilon C} \cdots \bigwedge_{x_n \varepsilon C} [\Phi \rightarrow \mathrm{Rel}(C,\Phi)]$$

(dual to the previous lemma).

__Corollary__: If $\Phi(x_1,\ldots,x_n)$ is Δ_1^{ZF} and C a transitive class, then

$$ZF \vdash \bigwedge_{x_1 \varepsilon C} \cdots \bigwedge_{x_n \varepsilon C} [\Phi \leftrightarrow \mathrm{Rel}(C,\Phi)].$$

__Absoluteness-lemma for L__: $ZF \vdash \bigwedge_x [x \varepsilon L \leftrightarrow \mathrm{Rel}(L, x \varepsilon L)]$.

__Proof.__ The formula $x\varepsilon L$ is defined to be $\bigvee_y (\mathrm{Ord}(y) \wedge x\varepsilon L(y))$. Since $\mathrm{Ord}(y)$ is Σ_0^{ZF} and $x \varepsilon L(y)$ is Δ_1^{ZF} both formulae are L-absolute according to the corollary, resp. a remark made above. Hence the relativization of $x\varepsilon L$ to L reduces to $\bigvee_y (\mathrm{Ord}(y) \wedge x\varepsilon L(y) \wedge y \varepsilon L)$ under the assumption $x\varepsilon L$. But since $ZF \vdash \bigwedge_y (\mathrm{Ord}(y) \rightarrow y\varepsilon L)$ we see that $\mathrm{Rel}(L, x\varepsilon L) \leftrightarrow x\varepsilon L$, q.e.d.

__Corollary__: The axiom V = L holds in the submodel $\langle \bigcup_\alpha M_\alpha, \in \rangle$. This follows, since the predicate "x is constructible" is absolute with respect to the class $L = \cup M_\alpha$.

This finishes the proof for Gödel's consistency theorem. Gödel's original proof for $P(\omega_\alpha) \cap L \subseteq L(\omega_{\alpha+1})$ (see [24] p. 221) was by direct computation that the order of any constructible subset m of ω_α is strictly less than $\omega_{\alpha+1}$ (using some cardinality arguments). The difficult proof of Gödel has been simplified (and clarified) later by using the Löwenheim-Skolem theorem (for such a proof see either P.J. COHEN's book: "Set theory and the Continuum Hypothesis"; Benjamin.New York-Amsterdam 1967, p. 95 ff, or the book "Théorie axiomatique des Ensembles" by J.L. KRIVINE (Presses Univ. de France, Paris 1969, p. 115 ff.). I think that the most elegant proof is by using Lévy's theorem and the Σ_1^{ZF}-definability of L, which was discovered by R.B. JENSEN and C. KARP [42].

H) A Theorem of H. Putnam and a Lemma of G. Kreisel

Is there always a constructible set of order $\alpha+1$ which is not of order α, when α is less than ω_1? H. Putnam has answered this question in the negative, in his paper:

[69] H. PUTNAM: A note on constructible sets of integers; Notre Dame J. of Formal Logic, vol. 4 (1963) p. 270-273.

It is clear that there is always a constructible set of order $\alpha+1$ which is not of order α, for M_α itself is a member of $M_{\alpha+1}$ but not of M_α. But there is not always a set of integers in $M_{\alpha+1}$ which is not already in M_α. In fact, Putnam shows that it can happen, that "for a long time" (a Δ_2^1-ordinal) we get no new sets of integers in the hierarchy of constructible sets, and then "pop!" a new set of integers appears. In order to state the results of Putnam we have to explain the notation ω_1^L. We know, that L contains all ordinals, but we do not know whether ω_1 is absolute (ω_0 is certainly absolute, since $x=\omega_0$ is Σ_0^{ZF}) and hence the first uncountable ordinal in the sense of L, denoted by ω_1^L, may be smaller than the first uncountable ordinal in the sense of the whole universe V.

Theorem (H. Putnam) (in ZF): These are ordinals $\alpha < \omega_1^L$, such that there is no set of integers in $L(\alpha+1)-L(\alpha)$.

Proof: Consider the structure $\langle L(\omega_1+2),\varepsilon\rangle$ and let T be the set of sentences true in this structure. By Gödel's theorem $P(\omega_0) \cap L \subseteq L(\omega_1)$, hence every constructible set of integers is constructed at some level β for $\beta < \omega_1$. Thus $L(\omega_1+1)-L(\omega_1)$ does not contain any set of integers. Therefore the following statement (abbreviated by Φ):

$$\bigvee_\alpha [\mathrm{Ord}(\alpha) \wedge \neg\bigvee_x (x \subseteq \omega_0 \wedge x\varepsilon L(\alpha+1)-L(\alpha))]$$

holds in $\langle L(\omega_1+2),\varepsilon\rangle$, which is a substructure of $\langle L,\varepsilon\rangle$. Now apply the Löwenheim - Skolem theorem in L and get a substructure $\langle D,\varepsilon\rangle$ of $\langle L(\omega_1+2),\varepsilon\rangle$, which is countable in the sense of L and in which all sentences of T are valid. Hence in particular Φ holds in $\langle D,\varepsilon\rangle$. By Mostowski's lemma $\langle D,\varepsilon\rangle$ is isomorphic to a transitive set D^*. But $\langle D,\varepsilon\rangle \models \Phi \leftrightarrow \mathrm{Rel}(D,\Phi)$ and $\mathrm{Rel}(D,\Phi) \leftrightarrow \mathrm{Rel}(D^*,\Phi)$, since $\langle D,\varepsilon\rangle$ and $\langle D^*,\varepsilon\rangle$ are isomorphic. Thus we know that $\mathrm{Rel}(D^*,\Phi)$ holds. Write $\Phi \triangleq \bigvee_\alpha \Psi(\alpha)$, then $\mathrm{Rel}(D^*,\Phi)$ implies $\mathrm{Rel}(D^*,\Psi(\alpha))$ for some ordinal $\alpha \varepsilon D^*$. Notice that Φ is a Σ_1^{ZF}-formula and that $\Psi(\alpha)$ is Δ_1^{ZF}, hence Π_1^{ZF}. Thus $\mathrm{Rel}(D^*,\Phi)$ implies Φ by one of the preceding absoluteness-lemmata for transitive classes. Hence

$$\mathrm{Rel}(D^*,\Phi) \leftrightarrow \bigvee_{\alpha\varepsilon D^*} \mathrm{Rel}(D^*,\Psi(\alpha)) \rightarrow \bigvee_{\alpha\varepsilon D^*} \Psi(\alpha)$$

But D is countable in the sense of L and D^* (in L) isomorphic to D. We get that $\alpha\varepsilon D^* \rightarrow$ "α is countable in the sense of L", and we infer that these are ordinals $\alpha < \omega_1^L$ such that there is no set of integers in $L(\alpha+1)-L(\alpha)$, q.e.d.

H. Putnam has even obtained a stronger result. Let Δ_2^1 be $\Sigma_2^1 \cap \Pi_2^1$ in Kleene's analytical hierarchy. Notice that every Δ_2^1-ordinal is countable and $\leqslant \omega_1^L$ (Shoenfield).

Theorem (H. Putnam [69] p. 271): Let β be a Δ_2^1-ordinal. There is an ordinal $\alpha < \omega_1^L$ such that there are no sets of integers in $L(\alpha+\gamma)-L(\gamma)$ for any $\gamma \leqslant \beta$.

The proof is like the one given above but using $\alpha+\beta$ instead of $\alpha+1$ throughout. Notice that Δ_2^1-ordinals are absolute for transitive models. A set α is called a Δ_2^1-ordinal iff the ordertype of the ordinal α is a Δ_2^1-wellordering of a set of natural numbers.

So far we have discussed questions concerning the constructible hierarchy and the strength of the axiom of constructibility "V=L" and have shown, that "V=L" implies the (GCH). Now we pose the question whether V=L enables to prove in ZF more numbertheoretic theorems than one can prove in ZF. A numbertheoretic formula Φ is a ZF-formula in which all quantifiers are restricted to ω such that the corresponding unrestricted formula is a formula of Peano-arithmetic (see Feferman [15] p. 50, e.g.). Hence a numbertheoretic formula Φ contains besides variables, logical symbols and brackets only symbols for ordinal addition, ordinal multiplication and the successor relation. G. Kreisel has remarked that ZF and ZF+V=L have the same numbertheoretic theorems (G. KREISEL: Some uses of Metamathematics; British Journal f. Philosophy of Sci. 7 (1956/57) p. 161-173 (see p. 165), and G. KREISEL: La prédicativité, Bull. Soc. Math. France 88 (1960) p. 371-391 (see p. 386)).

Lemma:(G.Kreisel): Every numbertheoretic formula Φ which is provable in ZF + V = L is provable in ZF.

Proof. Suppose there is a numbertheoretic formula Φ such that $ZF + V = L \vdash \Phi$ but $ZF \nvdash \Phi$. Then $ZF + \neg \Phi$ is consistent and has a model $\mathfrak{M} \triangleq \langle M,R \rangle$. Since $\mathfrak{M}$ need not to be standard we cannot apply Gödel's theorem prima facie in order to obtain a submodel of ZF + V = L. But a careful inspection of the proof shows that in fact also $\mathfrak{M}$ has a submodel $\mathfrak{N} \triangleq \langle N,R \rangle$ of ZF + V = L such that N is a R-transitive subclass of M. Hence Φ holds in $\mathfrak{N}$. But Φ is Σ_1^{ZF} since ordinal-addition, multiplication etc are Δ_1^{ZF} and Φ has only restricted quantifiers. Hence by the R-transitivity of N: $Rel(\mathfrak{N},\Phi) \rightarrow Rel(\mathfrak{M},\Phi)$ (this is a formula in which quantifiers are restricted to $\mathfrak{N}$, respect to $\mathfrak{M}$ and ε is replaced by R). But $\mathfrak{N} \models \Phi$ implies $Rel(\mathfrak{N},\Phi)$. Thus $Rel(\mathfrak{M},\Phi)$ which in turn implies $\mathfrak{M} \models \Phi$, a contradiction to our assumption $\mathfrak{M} \models \neg \Phi$, q.e.d.

I) A Theorem of Shoenfield

Our basic question was whether (AC) and (GCH) are provable or refutable in ZF. Gödel's result shows that neither (AC) nor (GCH) can be refutable in ZF. We do not yet know wether (AC), (GCH) or V = L are provable or independent from ZF. Since this question is difficult to answer we shall ask whether all sets of integers are constructible or even more restrictive whether all $\underset{\sim}{\Pi}^1_n$ or $\underset{\sim}{\Sigma}^1_n$ sets of integers in Kleene's analytical hierarchy for $n \leq k$ are constructible. How large is the upper bound k? We shall show that every Σ^1_2 and every Π^1_2 set of integers is constructible (J.R.Shoenfield). This is the best possible result. The results of Rowbottom, Gaifman and Silver show that the existence of some large cardinals (e.g. measurable or Ramsey ones) imply the existence of nonconstructible sets of integers. R.M. Solovay proved that in the set theory with large cardinals the set $0^{\#}$ of Gödel-numbers of sentences true in L (which can be defined here by results of J.Silver) is a nonconstructible, Δ^1_3-set of natural numbers, see: R.Solovay: A nonconstructible Δ^1_3-set of integers; Trans. Amer.Math.Soc., vol.127(1967)p.50-75. - Shoenfield' result, which we are going to prove, is contained in:

[76] J.R.SHOENFIELD: The problem of predicativity; In: Essays on the Foundations of Math's, dedicated to A.Fraenkel, Jerusalem 1962, p.132 - 139.

Definition: a set of integers or a set of sets of integers is $\underset{\sim}{\Sigma}^1_n$ ($\underset{\sim}{\Pi}^1_n$ respectively) iff X is definable in the standard model of the 2-nd order arithmetic by a formula in prenex form having n alternating 2-nd order quantifiers beginning with an existential (universal, resp.) one. In the defining formula are parameters allowed. (see: [75] p. 173 ff.)

Theorem (Shoenfield): If $\Phi(\alpha,\beta)$ is $\underset{\sim}{\Pi}^1_1$-predicate, β_0 a constructible function (from ω to ω) and $K = \{\alpha;\ \Phi(\alpha,\beta_0)\} \neq \emptyset$, then $K \cap L \neq \emptyset$, if the parameters lie in L.

Outline of proof: By a result of Kleene (Amer.J.Math.77(1955),p.405-428, see § 26) there is a number e such that

$$(*) \qquad \Phi(\alpha,\beta) \leftrightarrow (R^{\alpha,\beta} \text{ is a well-ordering})$$

where $R^{\alpha,\beta}$ is the predicate which is recursive in α,β with Gödel-number e. Since arithmetical predicates are absolute and since "x is a well-ordering" is a Δ^{ZF}_1-formula, we conclude that $\Phi(\alpha,\beta)$ is absolut with respect to the transitive class L. Shoenfield shows that $\bigvee_\alpha \Phi(\alpha,\beta)$ is also absolut. Then we have for constructible β_0:

$$\bigvee_\alpha \Phi(\alpha,\beta_0) \leftrightarrow \mathrm{Rel}(L,\bigvee_\alpha \Phi(\alpha,\beta_0)) \leftrightarrow \bigvee_{\alpha\in L} \mathrm{Rel}(L;\Phi) \leftrightarrow \bigvee_{\alpha\in L} \Phi(\alpha,\beta_0\ ).$$

This proves the theorem.

Corollary: If $\Phi(\alpha,\beta,\gamma)$ is an arithmetical predicate such that all parameters in Φ lie in L ,and γ_0 is constructible, then

(1) $\bigvee_\alpha \bigwedge_\beta \Phi(\alpha,\beta,\gamma_0) \leftrightarrow \bigvee_{\alpha\in L} \bigwedge_{\beta\in L} \Phi(\alpha,\beta,\gamma_0)$

(2) $\bigwedge_\alpha \bigvee_\beta \Phi(\alpha,\beta,\gamma_0) \leftrightarrow \bigwedge_{\alpha\in L} \bigvee_{\beta\in L} \Phi(\alpha,\beta,\gamma_0)$.

Proof. Using the theorem, $\bigvee_\alpha \bigwedge_\beta \Phi(\alpha,\beta,\gamma_0) \rightarrow \bigvee_{\alpha\in L} \bigwedge_\beta \Phi(\alpha,\beta,\gamma_0) \rightarrow \bigvee_{\alpha\in L} \bigwedge_{\beta\in L} \Phi(\alpha,\beta,\gamma_0)$. Again using the theorem, we have for α_0 c n-structible:

$$\bigvee_\beta \neg\, \Phi(\alpha_0,\beta,\gamma_0) \rightarrow \bigvee_{\beta\in L} \neg\, \Phi(\alpha_0,\beta,\gamma_0).$$

Contraposing and adding a quantifier $\bigvee_{\alpha\in L}$, we get

$$\bigvee_{\alpha\in L} \bigwedge_{\beta\in L} \Phi(\alpha,\beta,\gamma_0) \rightarrow \bigvee_{\alpha\in L} \bigwedge_\beta \Phi(\alpha,\beta,\gamma_0) \rightarrow \bigvee_\alpha \bigwedge_\beta \Phi(\alpha,\beta,\gamma_0).$$

This proves (1), and (2) follows by duality.

Theorem (Shoenfield): Every Σ^1_2- and every Π^1_2-subset of ω is a constructible set.

Proof: It follows from (1) that if $\Phi(\alpha,\beta,x)$ is arithmetical, then

$$\{x;\ \bigvee_\alpha \bigwedge_\beta \Phi(\alpha,\beta,x)\} = \{x;\ \bigvee_{\alpha\in L} \bigwedge_{\beta\in L} \Phi(\alpha,\beta,x)\} \in L(\zeta)$$

for some ordinal ζ, computed as in the proof for the replacementaxiom in L. Analoguously the theorem follows for Π^1_2-sets from (2) of the corollary, q.e.d.

Remark. Using the Kondo-Addison theorem Shoenfield has furthermore shown that the Δ^1_2-sets of integers are just the constructible sets of integers whose orders are Δ^1_2-ordinals. Another characterisation of Δ^1_2-sets is contained in B.Scarpellini, Transactions AMS 117(1965) p.441-450.

J) REMARKS

P.J.Cohen has shown that, if there are transitive standard-models of ZF, then there is a minimal such model (see his paper: A minimal model for set theory; Bull.Amer.Math.Soc.69(1963)p.537-540, see Mostowski's review in the Math.Reviews 27(1964)# 41,p.9). The main consequence is that if Φ is a ZF-formula whose consistency with ZF is provable by means of the minimal model or Gödels constructible model, then there is no ZF-formula $\Gamma(x)$ defining a transitive standard ZF-model such that ZF $\vdash$ Rel(G, $\neg\,\Phi$) if G = $\{x;\ \Gamma(x)\}$. Thus the method of transitive inner models will not give a proof for the independence of the (AC) or the (GCH) or "V = L". For a nice presentation of this result of P.J.Cohen see also C.Karp[42]p.18-21, and the paper:

[65] A.MOSTOWSKI: On models of ZF-Set Theory satisfying the axiom of Constructibility; Acta Philosophica Fennica, Fasc.18 (1965)p.135-144.

A proof for ZF + V = L $\vdash$ "there is a $\underset{\sim}{\Delta}^1_2$-set which has not the property of Baire"(theorem announced by Gödel in [23]) has been published by Addison:

[2] J.W.ADDISON: Some consequences of the axiom of constructibility; Math.46(1959)p.337-357.

Another interesting paper in this area is: V.A.LJUBECKII: Some corollaries of the hypothesis of noncountability of the set of constructible real numbers (Dokl.Akad.Nauk.SSSR - translation in: Soviet Math.Dokl.vol.9(1968),p.1195-1196).

Additions to chapter I

1) The result, that the schema of foundation (VII_Φ) does not follow from the axiom of foundation (VII) in Zermelo's set theory Z has been obtained independently by R.B.Jensen and Schröder (see p.17) and by M.Boffa: Axiome et schéma de fondement dans le système de Zermelo; Bull.Acad.Polon.Sci, vol.17(1969)p.113-115.

2) McNaughton (Proc.Amer.Math.Soc.5(1954)p.505-509) has proved that neither Zermelo's set theory Z nor Zermelo-Fraenkel's set theory ZF are finitely axiomatizable. Montague has extended McNaughton's result by proving that neither ZF nor any consistent extension T of ZF without new constants is a finite extension of Z (see p.20). R.Montague has first announced his result in abstract: Zermelo-Fraenkel set theory is not a finite extension of Zermelo-set theory; Bull.Amer.Math.Soc.62(1956)p.260. We have given reference to a paper of K.Hauschild in which McNaughton's result has been obtained by different methods. There is a further paper by K.Hauschild in which the non-finite axiomatizability of ZF is proved:

   K.Hauschild: Das Zermelo-Fraenkel'sche Axiomensystem der Mengenlehre; Wiss.Zeitschr.der Humboldt Univ.Berlin,Math.Nat.Reihe, vol.15(1966)p.1-3.

   In this paper Hauschild shows that every finite subsystem of ZF possesses countable models the domain of which can be denumerated by a formula belonging to the formalism of ZF.

Additions to chapter II

A modification of Gödel's consistency proof [25] has been published by L.Rieger:

L.Rieger: On the Consistency of the generalized continuum hypothesis. Rozprawy Matematyczne, vol.31 Warszawa 1963 (45 pages), Polska Akademia Nauk.

CHAPTER III

FRAENKEL - MOSTOWSKI - SPECKER MODELS

This chapter is devoted to the study of the socalled "Permutation Models". They provide a method for obtaining independence results. It was for the purpose of obtaining the relative consistency of $\neg$(AC) and negations of some weakened forms of (AC) that A.A.Fraenkel first introduced the notion of a permutation model:

[20] A.FRAENKEL: Der Begriff "definit" und die Unabhängigkeit des Auswahlaxioms; Sitzungsberichte d. Preussischen Akad. Wiss.,Phys.Math.Klasse,vol.21(1922)p.253-257.

[21] A.FRAENKEL: Ueber eine abgeschwächte Fassung des Auswahlaxioms; J.S.L. 2(1937)p.1-27. [Abstracts: C.R.Acad.Sci.Paris vol. 192(1931)p.1072, and: Jahresberichte der DMV 41(1931) part 2, p.88].

At that time (1922) the distinction between Mathematics and Metamathematics was not yet clear enough and Fraenkel's proof fails in so far as questions concerning the absoluteness of certain notions are not treated. This failing has been observed and corrected by A.Mostowski, partly in collaboration with A.Lindenbaum:

[54] A.LINDENBAUM - A.MOSTOWSKI: Ueber die Unabhängigkeit des Auswahlaxioms und einiger seiner Folgerungen; C.R.Soc.Sci. Lettr.Varsovie, Classe III, vol. 31(1938)p.27-32.

[64] A.MOSTOWSKI: Ueber die Unabhängigkeit des Wohlordnungssatzes vom Ordnungsprinzip; Fund.Math.32(1939)p.201-252.

Fraenkel has acknowledged the corrections and refinements obtained by Mostowski (see Fraenkel's review of [54] in the J.S.L.4(1939)p. 30-31). The major deficiency of the Fraenkel-Mostowski method is that it does not apply to "ordinary" set theories, id est, to set theories in which all the elements are sets and in which the axiom of foundations holds. The Fraenkel-Mostowski method is applied to set theories which admit the existence of "Urelements", id est objects which are not sets (have no elements and are distinct from the empty set $\emptyset$). These set theories are obtained from "ordinary" set theories (e.g. ZF) what essentially amounts to a weakening of the extensionality and dropping the axiom of foundation. Such a set theory, if obtained from ZF, will be called ZFU (the "U" indicating: with urelements). The FM-method has been ameliorated by E.P.Specker. His method, the socalled FMS-method (Fraenkel-Mostowski-Specker method), applies to set theories which permit the existence of unfounded sets, id est, they permit the existence of sequences of the type

$$.....\varepsilon\ x_n\ \varepsilon....\varepsilon\ x_3\ \varepsilon\ x_2\ \varepsilon\ x_1 \quad (n\ \varepsilon\ \omega)$$

or other similar phenomena such as $x = \{x\}$:

[82] E.SPECKER: Zur Axiomatik der Mengenlehre (Fundierungs- und Auswahlaxiom); Zeitschr.f.math.Logik und Grundlagen d. Math.vol.3(1957)p.173-210.

The set theories to which the FMS-method applies are obtained e.g. from ZF by weakening the axiom of foundation only. For a discussion of these methods we refer our reader to the book "Foundations of Set Theory" by A.Fraenkel - Y.Bar-Hillel (Amsterdam 1958)p.49-54, and to

[49] A.LÉVY: The Fraenkel-Mostowski Method for Independence Proofs in Set Theory; in: Symposium on the Theory of Models (Amsterdam 1965)p.221-228.

In the sequel we present Specker's method. We start with a proof that the existence of "reflexive" sets $x = \{x\}$ is relative consistent.

A) THE INDEPENDENCE OF THE AXIOM OF FOUNDATION

The notion of a well-founded set (ensemble ordinaire) was first considered by D.Mirimanoff (L'Ens.Math.19(1917)p.37-52,p.209-217, and vol.21(1920)p.29-52). A.Fraenkel (Math.Ann.86(1922)p.230-237) has added to the axioms of Zermelo an axiom of restriction (Axiom der Beschränktheit) asserting that every set is well-founded but the formulation was unfortunately not within the ZF-formalism (Frankel formulated it in a way similar to Hilbert's Vollständigkeitsaxiom of the "Grundlagen der Geometrie", 1899, see also Fraenkels paper in the Journal f.Math.141(1911) p.76).
It was Johann von Neumann (1925) who replaced Fraenkel's "axiom" by the axiom: $x \neq \emptyset \rightarrow \bigvee_y (y\ \varepsilon\ x \wedge x \cap y = \emptyset)$, (axiom(VII) in our list of ZF-axioms). J.v.Neumann's axiom excludes the existence of "extraordinaire" sets as it was the aim of Fraenkel's axiom, and is formulated within the ZF-formalism . Later Zermelo independently introduced this axiom and introduced the name "Axiom der Fundierung" for it (Zermelo, Fund.Math.16(1930) see p.31). The consistency of

the axiom of fundierung (VII) with ZF^0 was proved by J.v.Neumann. The consistency of $\neg$ (VII) with ZF^0 was first proved (independently, using different methods) by P.Bernays [J.S.L. 19(1954) see p.83-84 - this result was announced by Bernays already in 1941(J.S.L. 6, see p.10] and by E.P.Specker in his Habilitationsschrift, Zürich 1951, see [82]. The method for obtaining the indepence of the Fundierungsaxiom (VII) has been modified and simplified in the last 14 years by many authors [e.g. L.Rieger: Czechoslovak Math.J.7(1957)p.323-357; P.Hájek: Zeitschr.math.Logik u.Gr.Math.11(1965)p.103-115; M.Boffa: Zeitschr.math.Logik u.Gr.Math.14(1968)p.329-334]. A number of nice independence results is contained in:

[4] M.BOFFA: Les ensembles extraordinaires; Bull.Soc.Math.Belgique, vol.20(1968)p.3-15.

We shall present Bernays' independence proof but with the modifications due to L.Rieger and M.Boffa.

Permutations of the universe. Let $\Phi(x,y)$ be a ZF-formula having no free variables except x and y. Suppose Φ satisfies the following conditions:

(P.1) $ZF \vdash \bigwedge_x \bigwedge_y \bigwedge_z [\Phi(x,y) \wedge \Phi(x,z) \rightarrow y = z]$

(P.2) $ZF \vdash \bigwedge_x \bigwedge_y \bigwedge_z [\Phi(x,y) \wedge \Phi(z,y) \rightarrow x = z]$

(P.3) $ZF \vdash \bigwedge_x \bigvee_y \Phi(x,y) \wedge \bigwedge_y \bigvee_x \Phi(x,y)$.

Then Φ is said to define a permutation of the universe V. In this case define the class-term $F = \{\langle x,y \rangle ; \Phi(x,y)\}$ and $\langle x,y \rangle \varepsilon F \leftrightarrow y = F(x)$. Define a binary predicate ϵ_F by

$$x \epsilon_F y \underset{\text{Def.}}{\leftrightarrow} F(x) \varepsilon y \leftrightarrow \bigvee_z [\Phi(x,z) \wedge z \varepsilon y].$$

For a ZF-formula Ψ let $Rep(F,\Psi)$ be the formula obtained from Ψ by replacing the symbol ε by ϵ_F at all places of occurrence.

Theorem 1. Suppose $F = \{\langle x,y \rangle ; \Phi(x,y)\}$ defines a permutation of the universe, then $Rep(F,-)$ is a syntactic model of ZF^0 in ZF, id est $ZF \vdash Rep(F,\Psi)$ for every axiom Ψ of ZF^0.

Proof: Ad(0): The real empty set $\emptyset = F(a)$ satisfies the requirements of the relativized axiom of Null-set. Ad(I): suppose that $z \epsilon_F x \leftrightarrow z \epsilon_F y$ for all z. We prove $x \subseteq y$: Take $a \varepsilon x$. By (P.3) $a = F(b)$ for some b. Hence $F(b) \varepsilon x$, i.e. $b \epsilon_F x$. The hypothesis yields $b \epsilon_F y$, hence $a = F(b) \varepsilon y$. The part $y \subseteq x$ is proved analogously. By the axiom of extensionality of ZF we get $x = y$ as desired.

<u>Ad</u>(II): Let x and y be given. Put $z = \{F(x),F(y)\}$. Then $u \in_F z \leftrightarrow F(u) \varepsilon z$. Hence either $F(u) = F(x)$ or $F(u) = F(y)$. But the latter holds by (P.2) iff either $u = x$ or $u = y$. <u>Ad</u>(III): Let x be given and put $y = \{F(z);\ \bigvee_u (F(z) \varepsilon u \wedge F(u) \varepsilon x)\}$; then $z \in_F y \leftrightarrow \bigvee_u (z \in_F u \wedge u \in_F x)$ as required. The set y exists by virtue of the axioms of union and replacement of ZF. <u>Ad</u>(IV): Define (in ZF) a function f with domain ω by: $f(0) = \emptyset$, $f(1) = \{F(f(0))\}$, $f(n+1) = f(n) \cup \{F(f(n))\}$. Finally, define $x = \{F(f(i));\ i \varepsilon \omega\}$. We have that $z \in_F x \leftrightarrow z = f(i)$ for some i and $f(i) \subset_F f(i+1) = f(i) \cup_F \{f(i)\}_F$, where $\subset_F$ expresses proper F-inclusion i.e. "Inclusion" with respect to $\in_F$). Hence x satisfies the requirements of the relativized axiom of infinity.
<u>Ad</u>(V): For a set x the power-set in the sense of the model is the set $y = \{F(t);\ t \subseteq x\}$. In fact: $z \in_F y \leftrightarrow z \subseteq x$ (by (P.2)), therefore: $u \in_F z \in_F y \rightarrow F(u) \varepsilon z \wedge z \subseteq x$. Hence $F(u) \varepsilon x$, i.e. $u \in_F x$. <u>Ad</u>(VI): Let $\Gamma(x,y)$ be a ZF-formula with precisely two free variables x and y and let $\Psi(x,y)$ be $Rep(F,\Gamma(x,y))$. Suppose that for every x there is precisely one y such that $\Psi(x,y)$. Let a be any set. Define $a^* = \{F^{-1}(x);\ x \varepsilon a\}$. By the axiom of replacement (of ZF) corresponding to Ψ and a^* there is a set b^* such that

$$\bigwedge_y [y \varepsilon b^* \leftrightarrow \bigvee_u (u \varepsilon a^* \wedge \Psi(u,y))]$$

Define $b = \{F(y);\ y \varepsilon b^*\}$, then b satisfies the relativized replacement axiom corresponding to Ψ and a. Thus the theorem is proved.

<u>Theorem 2</u>: Suppose $F = \{\langle x,y \rangle;\ \Phi(x,y)\}$ defines a permutation of the universe V. If $\langle V,\varepsilon \rangle$ satisfies the axiom of choice, then $\langle V,\in_F \rangle$ satisfies the axiom of choice too, id est: $ZF + (AC) \vdash Rep(F, (AC))$.

<u>Proof</u>: Consider a set s such that s is "not empty" with respect to $\in_F$ and such that the F-elements of s are not F-empty and pairwise F-disjoint (the suffix "F" in front of a notion N indicates that the corresponding "notion in the sense of the model" is meant, id est: Rep(F,N)). Define

$$a^* = \{\{x;\ F(x) \varepsilon y\};\ F(y) \varepsilon a\}$$

a^* is a set of non-empty, pairwise disjoint sets (with respect to ε). By (AC) there is a set b^* such that: $b^* \cap y$ is a singleton for each $y \varepsilon a^*$. Now $b = \{F(x);\ x \varepsilon b^*\}$ is a F-choice set for a, i.e. satisfies the condition:

$\bigwedge_y [y \in_F a \to \bigvee^1_z (z \in_F b \wedge z \in_F y)]$.

This proves theorem 2.

Definition. A set x is called reflexiv iff $x = \{x\}$, id est iff the condition $\bigwedge_y (y \in x \leftrightarrow y = x)$ holds. The existence of reflexive sets contradicts the axiom of Fundierung.

Theorem 3. If ZF^0 is consistent, then $ZF^0 + \bigvee^1_x (x = \{x\})$ is also consistent.

Proof. Let $\Phi(x,y)$ be the formula $(x = 0 \wedge y = 1) \vee (x = 1 \wedge y = 0) \vee (x \neq 0 \wedge x \neq 1 \wedge x = y)$. Hence Φ defines a permutation of the universe such that only the ordinals 0 and 1 are interchanged. Let F be the one-to-one function $\{\langle x,y\rangle ; \Phi(x,y)\}$. Then $F(1) = 0 \in 1 = F(0)$, hence $1 \in_F 1$ and $1 = \{1\}_F$. Hence Rep(F,-) is a syntactic model of $ZF^0 + \bigvee^1_x x = \{x\}$ in ZF. The relative consistency follows from a theorem in chapt.I, section D. page 9, q.e.d.

Corollary: If ZF is consistent, then ZF^0 + "negation of the axiom of foundation" is consistent and ZF^0 + (AC) + "negation of the axiom of foundation" is consistent too.

This follows from theorems 2 and 3 and Gödel's consistency result. Hence the axiom of foundation is independent from ZF^0 + (AC).

Our next question is whether the axiom of foundation can be violated in such a form where there exists a countable set of reflexive sets or where there exists a proper class of reflexive sets which is in one-to-one correspondence with the class of all ordinals Both questions have been answered in the positive way by E.P.Specker (see also M.Boffa, Zeitschr.math.Logik u.Gr.d.Math. 14(1968)p.329-334 and [4]).

Theorem 4: If ZF is consistent, then ZF^0 + "there exists a set R of reflexive sets such that R is equipotent with ω" is consistent too.

Proof: Consider the following permutation F of the universe: $F(x) = \{x\}$ iff $x \in \omega^*$, $F(x) = y$ iff $x = \{y\}$ for $y \in \omega^*$ and $F(x) = x$ in all other cases; here ω^* is difined to be $\omega - \{1\} = \{0,2,3,4,\ldots\}$. For $x \in \omega^*$, $\{x\}$ is never in ω^* and F is well-defined. In particular we have: $F(0) = 1$, $F(1) = 0$, $F(2) = \{2\}$, $F(\{2\}) = 2$, etc. Hence $1 \in_F 1$ (since $F(1) = 0 \in 1$), id est: $\{0\} \in_F \{0\}$. Further $\{2\} \in_F \{2\}$, $\{3\} \in_F \{3\}$, etc. and obviously $F(x) \in \omega^* \to x = \{F(x)\} = \{x\}_F$.

We shall show that there is a F-correspondence between the F-set of reflexive sets $\mathbf{n} = \{n\}_F$ for $n \geq 1$ and ω.

The unordered pair in the model-sense of x and y is $\{F(x), F(y)\}$. The ordered pair in the sense of the model is hence

$$\langle x,y \rangle_F = \{F(\{F(x)\}), F(\{F(x), F(y)\})\}$$

The natural numbers in the sense of the model are the sets $f(0) = \emptyset$, $f(n+1) = f(n) \cup \{F(f(n))\}$ (see the proof of th.1). The function

$$g = \{\langle f(n). \{n+2\}\rangle_F;\ n \in \omega\}_F = \{F(\langle f(n), \{n+2\}\rangle_F;\ n \in \omega\}$$

is a function in the F-sense between $\omega_F = \{f(n);\ n \in \omega\}_F = \{F(f(n));\ n \in \omega\}$, the F-set of F-natural numbers, and some F-reflexive sets $\{n+2\} = \{F(\{n+2\})\}$ for $n \in \omega$.

Corollary: If ZF is consistent, then ZF^0 + "there exists a proper class **R** of reflexive sets such that there is a one-to-one correspondence between the class of all ordinals and R" is consistent too.

Proof. Consider the following permutation of the universe: $F(x) = \{x\}$ iff $x \in On^*$, $F(x) = y$ iff $x = \{y\}$ for $y \in On^*$ and $F(x) = x$ otherwise. Here On is the class of all ordinal-numbers and $On^* = On - \{1\}$. Now proceed as in the proof of th.4.

Remark. By theorem 2 it is possible to add (AC) to ZF^0 in th.4 and its corollary above. Further, in the model of theorem 3 the (GCH) holds. Hence the axiom of foundation is indepentent from ZF^0 + (AC) + (GCH). This cannot be strengthened by adding V = L, since obviously the axiom of foundation follows from V = L. For further consistency results like those proved in theorems 3 and 4 and its corollary consult the papers of M.Boffa (those already cited and Boffa's papers in the C.R.Acad.Sc.Paris vol.264(1967)p.221-222,vol.265(1967) p.205-206, vol.266(1968)p.545-546, vol.268(1969)p.205). In particular Boffa's second C.R.-paper contains the following fine result:

Theorem (M.Boffa): Let $\langle s,< \rangle$ be any partially ordered set. If ZF is consistent, then so is ZF^0 + (AC) + "there is a transitive set t such that $\langle s,< \rangle$ and $\langle t,\epsilon \rangle$ are isomorphic".

In particular $\langle s,< \rangle$ can be taken to be any linearly ordered set (e.g. a dense totally ordered set).

It is known that in ZF + (AC) + "There are strongly inaccessible cardinal numbers "the class of Grothendiek-universa is totally ordered by $\subseteq$ (see G.Sabbagh, Archiv d. Math. 20(1969)p.449-

456). U.Felgner has shown that it is consistent with ZF^0 + (AC) + $\bigwedge_x \bigvee_y (\bar{\bar{x}} < \bar{\bar{y}} \wedge In(\bar{\bar{y}}))$ that given any partially ordered set $\langle s, \leq \rangle$ there is a set t of Grothendiek-universa, such that $\langle s, \leq \rangle$ and $\langle t, \subseteq \rangle$ are isomorphic (see U.Fg.Archiv d.Math.20(1969)p.561-566: [17], and M.Boffa - G.Sabbagh [5] .

B) THE FRAENKEL-MOSTOWSKI-SPECKER METHOD

A weak axiom of foundation, compatible with the existence of reflexive sets.

Let A be a set of reflexive sets. Define $R_0(A) = A$, $R_\alpha(A) = \bigcup\{P(R_\beta(A)); \beta < \alpha\}$ for $\alpha > 0$, and $W(A) = \bigcup\{R_\alpha(A); \alpha \in On\}$. One proves that $\alpha \leq \beta \rightarrow R_\alpha(A) \subseteq R_\beta(A)$ and that all sets $R_\alpha(A)$ are transitiv. It is not provable that $\bigvee_x (V = W(x))$ (see the corollary in the preceeding section!) but it is consistent with ZF^0 (same proof as in the consistency proof for $V = \bigcup V_\alpha$). In W(A) it holds that every set is wellfounded relative to A:

(WF) Axiom of weak foundation: $\bigvee_A (\bigwedge_x x \neq \emptyset \rightarrow \bigvee_y (y \in x \wedge (y \cap x = \emptyset \vee A \ni y = \{y\})))$.

All models considered from now on (in this chapter) will satisfy this axiom. From now on we assume the axiom $\bigvee_x (V = W(x)) \wedge \bigwedge_y (y \in x \rightarrow y = \{y\}))$.

Automorphisms of the universe. An automorphism of V is a one-to-one mapping τ from V onto V such that $x \in y \leftrightarrow \tau(x) \in \tau(y)$.

Let A be a basis of reflexive sets for V, id est V = W(A), and let π be any permutation of A (i.e. one-to-one mapping from A onto A), then π can be extended in a unique way to an automorphism π^* of V. This is done by induction: suppose we have extended π so that π acts on all sets of $R_\alpha(A)$. Let $x \in R_{\alpha+1}(A)$ and define

$$\pi(x) = \{\pi(y); y \in x\}$$

The uniqueness follows from the fact that a notion of rank, $\rho(x)$, can be defined: $\rho(x) = Min\{\alpha; x \subseteq R_\alpha(A)\}$. This shows that every automorphism τ of V is uniquely determined by a permutation of A. If π permutes A and π^* is its extension, then

$$(\pi^{-1})^* = (\pi^*)^{-1} \quad \text{and} \quad (\pi_1\pi_2)^* = \pi_1^*\pi_2^*.$$

Hence the group of permutations of A and the automorphism group of V (written as $Aut(V,\varepsilon)$) are isomorphic and we need not to distinguish between them.

Filters of subgroups. If G is any (multiplicatively written) group, H a subgroup of G, and $g \in G$, then $g^{-1}Hg$ is called a conjugate subgroup of G, conjugate with respect to H.

Definition. A non-empty set F of subgroups of a group G is called a filter (of subgroups of G) iff the following three conditions hold:

(i) $H \varepsilon F \wedge g \varepsilon G \rightarrow g^{-1}Hg \varepsilon F$.

(ii) $H_1 \varepsilon F \wedge H_1 \leqslant H_2 \wedge H_2 \leqslant G \rightarrow H_2 \varepsilon F$.

(iii) $H_1 \varepsilon F \wedge H_2 \varepsilon F \rightarrow H_1 \cap H_2 \varepsilon F$.

Here $H_1 \leqslant H_2$ means that H_1 is a subgroup of H_2. We shall show that every filter F of subgroups of any subgroup G of Aut(V,ε) determines a model of ZF^0.

Definition of the model $\mathfrak{M}[G,F]$. Let $G \leqslant Aut(V,\varepsilon)$ and F a filter on G. For any set x let $H[x] = \{\tau \varepsilon G; \tau''x = x\}$ where $\tau''x = \{\tau(y); y \varepsilon x\}$. Obviously H[x] is a subgroup of G. Again let C(x) denote the transitive closure of x, i.e. $C(x) = \{x\} \cup x \cup \bigcup x \cup \ldots$

Now define $M = \{x; \bigwedge_y (y \varepsilon C(x) \rightarrow H[y] \varepsilon F)\}$.

Sets of the "model" $\mathfrak{M}[G,F]$ are thus elements of M and the membership-relation of $\mathfrak{M}[G,F]$ is the one of the whole universe V. We shall prove that

$$\mathfrak{M} \triangleq \mathfrak{M}[G,F] \triangleq \langle M,\varepsilon \rangle$$

is a model of ZF^0. But first we list some properties of $\langle M,\varepsilon \rangle$.

(α) M is a transitive class.

(β) If x is a subset of M, then x ε M iff H[x] ε F.

Theorem (Specker [82] p.196): (In $ZF^0 + V = W(A)$): $\mathfrak{M}[G,F] \triangleq \langle M,\varepsilon \rangle$ is a model of ZF^0.

Proof. Ad(0): Since $H[\emptyset] = G \varepsilon F$ and $\emptyset \subseteq M$, $\emptyset \varepsilon M$ and $\emptyset$ satisfies the axiom of Null-set in $\mathfrak{M}$.

Ad(I): The axiom of extensionality in $\mathfrak{M}$ follows from (α).

Ad(II): If x and y are sets of $\mathfrak{M}$, then H[x] ε F and H[y] ε F. Since F is a filter: $H[x] \cap H[y] \varepsilon F$. But $H[x] \cap H[y] \leqslant H[\{x,y\}]$, hence $H[\{x,y\}] \varepsilon F$. By (β) is {x,y} a set of $\mathfrak{M}$.

Ad(III): Let x be a set of $\mathfrak{M}$; hence H[x] ε F. We shall show that $H[x] \leqslant H[\bigcup x]$. $\bigcup x = \{z; \bigvee_y (z \varepsilon y \varepsilon x)\}$. For $\tau \varepsilon H[x]$: $z \varepsilon y \varepsilon x \rightarrow \tau(z) \varepsilon \tau(y) \varepsilon \tau(x)$ since τ is an automorphism. But $\tau \varepsilon H[x] \rightarrow \tau(x) = x$ and $\tau(y) = y' \varepsilon x$. Thus: $\bigvee_y (z \varepsilon y \varepsilon x) \leftrightarrow \bigvee_y (\tau(z) \varepsilon y \varepsilon x)$ for $\tau \varepsilon H[x]$. Hence $\tau(\bigcup x) = \tau\{z; \bigvee_y (z \varepsilon y \varepsilon x)\} = \{\tau(z); \bigvee_y (z \varepsilon y \varepsilon x)\} \subseteq \bigcup x$. But similar $\tau^{-1}(\bigcup x) \subseteq \bigcup x$ follows. Hence (if 1 denotes the identical mapping):

$$\bigcup x = 1(\bigcup x) = \tau\tau^{-1}(\bigcup x) \subseteq \tau(\bigcup x) \subseteq \bigcup x.$$

Thus $\tau \in H[\bigcup x]$. Now by (ii) of the filter-definition $H[\bigcup x] \in F$. By (β) $\bigcup x \in M$.

Ad(IV): By induction one shows that M contains all ordinals (using (II) and (III) already proved). Hence $\omega \in M$ and ω satisfies the requirements of the axiom of infinity.

Ad(V): Let x be any set. We shall show that $P(x) \cap M \in M$. If $y \subseteq x$ and $\tau \in H[x]$, then $\tau(y) \subseteq x$; hence $\tau(P(x)) \subseteq P(x)$. Moreover, if $y \subseteq x$ and $y \in M$, then $\tau(y) \in M$ (see the lemma 1 below) and therefore

$$\tau(P(x) \cap M) \subseteq P(x) \cap M.$$

The same holds for τ^{-1}. Hence $\tau(P(x) \cap M) = P(x) \cap M$ (as above) and $H[x] \subseteq H[P(x) \cap M]$. Thus $P(x) \cap M \in M$. The set $P(x) \cap M$ satisfies the requirements of the power-set áxiom relativized to $\mathfrak{M}$. Instead of proving directly that the replacement schema is true in the model we shall prove first that the Aussonderungsschema holds in it. Let $\Phi(x,x_1,\ldots,x_n)$ be a ZF-formula with no free variables other than $x,x_1,\ldots,x_n$. We have to show that if $x_1,\ldots,x_n y \in M$ then there is a set $z \in M$ such that

$$\bigwedge_x [x \in z \leftrightarrow x \in y \wedge Rel(M,\Phi(x,x_1,\ldots,x_n))]$$

First one shows by induction on the length of Φ that for every $\tau \in G$: $x,x_1,\ldots,x_n \in M \rightarrow (Rel(M,\Phi(x,x_1,\ldots,x_n)) \leftrightarrow Rel(M,\Phi(\tau(x),.,\tau(x_n))))$. In our present case, since $x_1,\ldots,x_n,y \in M$, thus $H[x_1],\ldots,H[x_n],H[y] \in F$, hence $H[x_1] \cap\ldots\cap H[x_n] \cap H[y] = H_0 \in F$, we can assume that all of $x_1,\ldots,x_n,y$ are H_0-symmetric (i.e. invariant under automorphisms $\tau \in H_0$). Consider the set

$$z = \{x;\ x \in y \wedge Rel(M,\Phi(x,x_1,\ldots,x_n))\}.$$

In order to show that $z \in M$ it is by (β) enough to prove that z is H_0-symmetric. Then $H_0 \leqslant H[z]$, hence by (ii) $H[z] \in F$ which implies by (β) that z is in M.

Hence take $\tau \in H_0$ and $x \in z$. Then $x \in y$ and $\tau(x) \in \tau(y)=y$ since $H_0 \leqslant H[y]$ by definition of H_0. Also, since $x \in z$ we have $Rel(M,\Phi(x,x_1,\ldots,x_n))$, hence $Rel(M,\Phi(\tau(x),\tau(x_1),\ldots,\tau(x_n))$. But since $\tau \in H_0 \leqslant H[x_i]$ $(1 \leqslant i \leqslant n)$, $\tau(x_i) = x_i$, and therefore $\tau(x) \in z$. Altogether: $\tau \in H_0 \rightarrow \tau(z) \subseteq z$. Hence $\tau(z) = z$ (proved as in (III)), and z is H_0-symmetric, q.e.d.

So far we have shown, that $\mathfrak{M} \simeq \mathfrak{M}[G,F]$ is a model of Zermelo-set theory Z. In order to prove that $\mathfrak{M}$ satisfies the replacement-axiom we need a lemma

Lemma 1. If $x \in M$ and $\tau \in G$, then $\tau(x) \in M$.

Proof by induction on the ε-relation.
$x \in M \rightarrow H[x] \in F$. We claim that

$$H[\tau(x)] \geq \tau H[x]\tau^{-1}.$$

Hence take $\sigma \in H[x]$. Then $(\tau\sigma\tau^{-1})\tau(x) = \tau\sigma(\tau^{-1}\tau)(x) = \tau\sigma(x) = \tau(x)$. Thus $\tau\sigma\tau^{-1} \in H[\tau(x)]$. It follows by (i) and (ii) of the filter definition that $H[\tau(x)] \in F$. Now, if x is reflexive (id est $x \in R_0(A)$), then $\tau(x)$ is by definition of the action of τ also in $R_0(A)$, hence reflexive. But $y \in C(\tau(x)) \rightarrow y = \tau(x)$. Hence $y \in C(\tau(x)) \rightarrow H[y] = H[\tau(x)] \in F$ and we get $x \in M$. If $x \in R_\alpha(A)$ and $y \in M \rightarrow \tau(y) \in M$ for all $y \in R_\beta(A)$ for $\beta < \alpha$, then $x \in M$ implies $x \subseteq M$, hence $\tau(x) \subseteq M$. Thus, by (β), $H[\tau(x)] \in F$ implies $\tau(x) \in M$, q.e.d.

Now we return to the proof of Specker's theorem. In the presence of the axiom schema of subsets (Aussonderung) the axiom schema of replacement is equivalent to the schema

$$\bigwedge_u \bigwedge_v \bigwedge_w [\Phi(u,v) \wedge \Phi(u,w) \rightarrow v = w] \rightarrow \bigwedge_y \bigvee_z \bigwedge_u \bigwedge_v (u \in y \wedge \Phi(u,v) \rightarrow v \in z).$$

where $\Phi(u,v)$ is a ZF-formula. Now let $\Phi(u,v)$ be such a formula with no free varibles other than $u,v,x_1,\ldots,x_n$. Assume that for $x_1,\ldots,x_n \in M$ and all $u,v,w \in M$ we have that $Rel(M,\Phi(u,v)) \wedge Rel(M,\Phi(u,w))$ implies $v = w$. Let y be a set, $y \in M$. Define

$$t = \{v \in M; \bigvee_u (u \in y \wedge Rel(M,\Phi(u,v)))\}$$

By the replacementaxiom in ZF, t is a set. Put $z = \bigcup\{\tau(t);\ \tau \in G\}$. Again by the replacementaxiom of ZF, z is a set since G is a set. Further $t \subseteq M$, hence $\tau(t) \subseteq M$, by lemma 1, thus $z \subseteq M$. Since G contains the identical mapping: $t \subseteq z$. Further z is G-symmetric, id est $\tau(z) = z$ for all $\tau \in G$, since as a group G is closed under products. Hence $H[z] = G \in F$, and by (β): $z \in M$. Thus the relativized weakened form of the replacement schema holds. This finishes the proof of Specker's theorem.

The model $\mathfrak{M}[G,F]$ was constructed relative to the set A of reflexive sets such that $\mathfrak{M}[G,F]$ is a subclass of $W(A) = \bigcup_\alpha R_\alpha(A)$. It is remarkable that neither A nor the elements of A are always sets in the model $\mathfrak{M}[G,F]$. But in all applications of the Fraenkel-Mostowski-Specker method we just want to have A as a set in $\mathfrak{M}$. This will be the case if the filter F satisfies the following additional condition:

(iv) $\bigwedge_x [x \in A \rightarrow H[x] \in F]$

It is easily seen (see the proof of lemma 1) that for $\mathfrak{M}[G,F] \triangleq (M,\varepsilon)$ and F satisfying (i), (ii)(iii)(iv), $A \subseteq M$, hence $A \in M$ (since $H[A] = G \in F$) by (β), holds.
Further , if F satisfies (i),...,(iv) then in $\mathfrak{M}[G,F]$ the weak axiom of foundation (WF) holds (if in the surrounding set-theory (WF) holds).

By definition $\tau \in H[x] \rightarrow \tau''x = x$, but τ need not to be the identical mapping on x. Define for any set x:

$$K[x] = \{\tau \in G;\ \tau \upharpoonright x = 1_x\}$$

where 1_x is the identical mapping on x and $\tau \upharpoonright x$ is the restriction of τ to x. Remark that always $K[x] \leqslant H[x] \leqslant G$. If $H[x] \in F$ then $K[x]$ need not to be in F, but if $K[x] \in F$ then there is a wellordering of x in $\mathfrak{M}[G,F]$, if the axiom of choice holds in the surrounding set theory.

Lemma 2: Every $\tau \in G$ acts as the identity on the well-founded part $M \cap \bigcup_\alpha V_\alpha$ of M.

Proof by induction on the Mirmanoff-rank $\rho(x)$ for well-founded sets x.

Lemma 3: (In ZF^0 + (WF) + (AC)): If $G \leqslant Aut(V,\varepsilon)$ and F is a filter of subgroups of G satisfying conditions (i), (ii), (iii), then $\mathfrak{M}[G,F]$ contains wellordering relations for each well-founded set x of $\mathfrak{M}[G,F]$. Hence the axiom of choice holds in the well-founded part of $\mathfrak{M}[G,F]$.

Proof. If x is well-founded, then by lemma 2: $H[x] = K[x] = G \in F$. Since the (AC) holds in the surrounding set theory x can be mapped one-to one on an ordinal α. But obviously (by lemma 2) every well-founded set of the surrounding set theory is contained in $\mathfrak{M}[G,F]$ (the well-founded sets of $\mathfrak{M}[G,F]$ are just the well-founded sets of the surrounding set theory!) and hence α is in $\mathfrak{M}[G,F]$ and the one-to-one mapping f from x onto α is also a well-founded set, hence also in $\mathfrak{M}[G,F]$.

Lemma 4: (In ZF^0 + (WF) + (AC)): If $G \leqslant Aut(V,\varepsilon)$ and F is a filter of subgroups of G satisfying (i), (ii) and (iii) then a set x of $\mathfrak{M}[G,F]$ can be mapped in $\mathfrak{M}[G,F]$ in a one-to-one fashion onto a well-founded set y of $\mathfrak{M}[G,F]$ iff $K[x] \in F$.

Proof. a) Suppose that there is such a one-to-one mapping f in $\mathfrak{M}[G,F]$ from $x \in M$ onto a well-founded set $y \in M$. Since $f \in M \to H[f] \in F$ it is sufficient to show that $H[f] \leqslant K[x]$ in order to verify that $K[x] \in F$ holds. Hence, take $\tau \in H[f]$ and $u \in x$. Then

$$(\bullet) \qquad \tau(\langle u,f(u)\rangle) = \langle \tau(u), \tau(f(u))\rangle = \langle \tau(u), f(u)\rangle$$

since $f(u) \in y$, hence $f(u)$ well-founded, hence $\tau(f(u)) = f(u)$ by lemma 2. But $\tau \in H[f] \to \tau(f) = f$, hence $\langle u,f(u)\rangle \in f \to \tau(\langle u,f(u)\rangle) \in f$, which means by ($\bullet$): $\langle \tau(u), f(u)\rangle \in f$. But f is one-to-one, hence $\tau(u) = u$. Thus $\tau \in K[x]$ and $H[f] \in F$, $H[f] \leqslant K[x]$ implies by (ii) $K[x] \in F$.

b) In order to prove the converse, suppose that for some $x \in M$ we have that $K[x] \in F$. Since $\mathfrak{M}[G,F] \triangleq \langle M,\varepsilon\rangle$ is a model of ZF^0 the axiom of pairing and union hold in it, and there are subsets s of x which can be mapped in $\mathfrak{M}$ one-to-one onto well-founded sets of $\mathfrak{M}$ (e.g. singletons). For each such map f_s, $s \subseteq x$, it holds that $K[x] \leqslant H(f_s)$. The set (in the sense of the surrounding set theory) of functions f_s, ordered by "is an extension of" is inductively ordered and has therefore by Zorn's lemma a maximal element f_0. The maximality of f_0 implies that f_0 must be defined on the whole set x (this argument takes place in the surrounding set theory ZF + (WF) + (AC)). But $f_1 \subseteq f_2 \to K[x] \leqslant H[f_2] \leqslant H[f_1]$. Hence $K[x] \leqslant H[f_0]$ and $H[f_0] \in F$. Since $Rg(f_0) \subseteq M$ and $Dom(f_0) = x \subseteq M$ we infer that $f_0 \subseteq M$. Hence, by (β), $f_0 \in M$, and lemma 4 is proved.

Corollary 5: (In ZF + (WF) + (AC)). A set x of the model $\mathfrak{M}[G,F]$ is well-orderable in $\mathfrak{M}[G,F]$ iff $K[x] \in F$.

Proof. If $K[x] \in F$, the there is by lemma 4 in $\mathfrak{M}$ a one-to-one mapping f from x onto a well-founded set y. By lemma 3, y can be well-ordered.Hence f induced a welloredering of x. If, conversely, x can be wellordered in $\mathfrak{M}$, then there is in $\mathfrak{M}$ a one-to-one mapping f from x onto an ordinal α where f and α are in $\mathfrak{M}$. But α is well-founded, hence $K[x] \in F$ by lemma 4.

C) THE INDEPENCENCE OF THE AXIOM OF CHOICE

In the set theory ZF^0 + (WF) the reflexive sets in the base $R_0(A) = A$ are all different but not too much distinguished. We shall use this fact in the construction of a ZF^0-model $\mathfrak{M}$ in which (AC) does not hold.

But in order to ensure that choice fails in $\mathfrak{M}$ we require that $\mathfrak{M}$ satisfies some symmetries. These symmetries of $\mathfrak{M}$ are determined by the group $G \leq Aut(V,\varepsilon)$ and the filter F of subgroups of G. Further suitable choices of groups G and filters F will give ZF^0-models in which the "general" axiom of choice fails but certain fragments of (AC) are true.

The Model of A.Fraenkel.

We shall present the model constructed by Fraenkel [21] but with the modifications due to E.P.Specker [82] (see [82] p.197-198). We work in a set theory ZF^0 + (AC) in which there is a countable set A of reflexive sets such that the weak axiom of foundation (WF) holds with respect to A, id est $V = \bigcup_\alpha R_\alpha(A)$. The consistency relative to ZF^0 was shown in section A).

Let $\{a_0, a_1, a_2, \ldots\} = A$ be a fixed enumeration of A. A transposition of A is a one-to one mapping π from A onto A which interchanges just two elements of A and is the identical mapping for all other elements. Call a transposition π of A "kind" iff π interchanges a_{2k} with a_{2k+1}. If $B = \{\{a_{2k}, a_{2k+1}\};\ k \in \omega\}$, then $\bigcup B = A$ and every kind transposition π maps B onto itself. Now let G be the group of those permutations τ of A which are a finite product of kind transpositions. Obviously G is abelian. Let F be the set of subgroups H of G whose index in G is finite:

$$F = \{H;\ H \leq G \wedge [G:H] < \aleph_0\}$$

We shall show that F is a filter satisfying conditions (i), (ii), (iii) and (iv).

Lemma (H.Poincaré): The intersection of a finite number of subgroups of finite index has finite index.

Proof. Let G be any group and let H_1 and H_2 be subgroups of G of finite index. Elements a and b of G lie in the same right coset of $H_1 \cap H_2$ iff $ab^{-1} \in H_1 \cap H_2$. Thus we obtain all right cosets of $H_1 \cap H_2$ by taking all non-empty intersections of right cosets of H_1 with right cosets of H_2. Thus $[G:H_1 \cap H_2]$ is finite. - The general case follows by induction on the number of subgroups.

It is well-known that $H_1 \leq H_2 \leq G$, then $[G:H_1] = [G:H_2]\cdot[H_2:H_1]$, and $H \leq G,\ g \in G \rightarrow [G:H] = [G:g^{-1}Hg]$. Hence our set F is in fact a filter, satisfying conditions (i), (ii), (iii). (For the grouptheoretical background we refer the reader to "Group Theory" by W.R.

SCOTT (Prentice-Hall, Inc. New Jersey 1964) p.20).

If $a \in A$, then $[G:H[a]] = 2$ (obviously), hence for all $a \in A$ the groups $H[a]$ are in F and F satisfies condition (iv). Thus the base A of reflexive sets is a set of the ZF^0-model $\mathfrak{M}[G,F] \triangleq \langle M,\varepsilon\rangle$. Now $B \subseteq M$ and since $H[B] = G \in F$ we infer that also the countable set B is a set of $\mathfrak{M}[G,F]$.

Lemma. In $\mathfrak{M}[G,F]$ the set $B = \{\{a_{2k},a_{2k+1}\};\ k \in \omega\}$ has no choice set.

Proof. Suppose there would be in $\mathfrak{M}[G,F]$ a choice set, say C, of B, such that C contains precisely one element from each set $\{a_{2k},a_{2k+1}\}$. Then $H[C]$ would be in F. But the only permutation $\tau \in G$ which fixes C is the identity. Hence $H[C]$ would be the trival subgroup $E = \{e\}$ when e is the identical mapping on A. But E has infinite index in G, a contradiction to $H[C] \in F$. B is countable in $\mathfrak{M}$, thus:

Corollary: The weak axiom of choice (AC_2^{ω}) saying, that every countable set of unordered pairs has a choice function, is independent from the axioms of the system ZF^0 + (WF).

A model of E. Specker.

Again we choose ZF^0 + (AC) + "there is a countable set A of reflexive sets such that $V = \bigcup_\alpha R_\alpha(A)$" as surrounding set theory. Let G be the group of those one-to-one mappings π from A onto A which move only finitely many elements of A. Define $F^* = \{K[t];$ t is a finite subset of $A\}$ and let F be the filter of subgroups of G such that F^* is a filter-basis of F. Specker shows in [82] p.198-199 that in the model $\mathfrak{M}[G,F]$ the following holds:

(1) The powerset of A is neither finite nor transfinite (i.e. contains no countable subset). The powerset of A is hence Dedekind-finite but not finite.
(2) There is no one-to-one mapping f from $A \times A$ into the powerset of A. Hence the statement $\bigwedge_m (m^2 \leqslant 2^m)$ fails in Specker's model.

D) THE INDEPENDENCE OF THE GENERALIZED CONTINUUM-HYPOTHESIS FROM THE ALEPH-HYPOTHESIS

We shall obtain the result mentioned in the heading by constructing Fraenkel's model in a set theory ZF^0 + (GCH) in which there exists a countable set A of reflexive sets such that $V = \bigcup_\alpha R_\alpha(A)$ is true. The following lemma shows that the latter theory is consistent and can thus be used to construct in it the model of Fraenkel.

Lemma. If ZF is consistent, then ZF^0 + (GCH) + "there exists a countable set A of reflexive sets such that $V = \bigcup_\alpha R_\alpha(A)$" is consistent too.

Proof. If ZF is consistent, then ZF + (GCH) is consistent by Gödel's theorem. Define a permutation F of the universe as in theorem 4 of chapter 3, section A. It was shown there (and in theorem 2), that

$$ZF + (AC) \vdash Rep(F, ZF^0) \wedge Rep(F,(AC))$$

Now let x be any set of power $\aleph_\alpha$ and let $P_F(x)$ be the power set of x in the sense of F, id est $P_F(x) = \{F(t);\ t \subseteq x\}$. Notice that x has also in the F-sense power $\aleph_\alpha$. But then x can be mapped one-to-one onto $\omega_\alpha - \omega_0 = \{\beta;\ \omega \leq \beta < \omega_\alpha\}$. We may assume that $x \cap (\omega \cup \{F(\gamma);\ \gamma \in \omega \wedge \gamma \neq 1\}$ is empty (otherwise map x onto such a set). Then the 1-1-mapping f from x onto $\omega_\alpha - \omega_0$ is not moved by F. By (GCH) $P(\omega_\alpha - \omega_0)$ can be mapped in a 1-1-fashion onto $\omega_{\alpha+1} - \omega_0$, hence the same holds for P(x). But in the model determined by F, $\omega_{\alpha+1}$ and $\omega_{\alpha+1} - \omega_0$ are equipotent and P(x) has in the F-sense power $\aleph_{\alpha+1}$. Thus

$$ZF + (GCH) \vdash Rep(F,(GCH)).$$

Let A be the set $\{F(x);\ F(x) \in \omega^*\} = \{x; F(x) \in \omega^*\}_F$.
This shows that ZF^0 + (GCH) + "there is a countable set A of reflexive sets" is consistent. But similar to v.Neumann's procedure we may restrict the universe of all sets to those which lie in some $R_\alpha(A)$ without violating any ZF^0-axiom nor the (GCH). The lemma is thus proved.

Now we proceed in the set theory whose consistency is assured by the lemma and define in it Fraenkel's model $\mathfrak{M}[G,F]$ as in section C). It was shown there that the set A is in $\mathfrak{M}[G,F]$ and that there is no wellordering of A in $\mathfrak{M}[G,F]$. Hence the (GCH) fails in Fraenkel's model. The technique for showing that the aleph-hypothesis (AH): $\bigwedge_\alpha (2^{\aleph_\alpha} = \aleph_{\alpha+1})$ holds in $\mathfrak{M}[G,F]$ is simply by proving that the model inherits the (AH) from the surrounding set theory.

Lemma. If the aleph-hypothesis holds in the surrounding set theory, then the aleph-hypothesis holds also in Fraenkel's model.

Proof. Let x be a set in Fraenkel's model $\mathfrak{M}[G,F] \triangleq \langle M,\varepsilon \rangle$ such that in $\mathfrak{M}$, x is wellordered and of power $\aleph_\alpha$. Then there is in $\mathfrak{M}$ a one-to-one function f from x onto the initial ordinal ω_α. But ω_α is a wellfounded set. Hence $K[x] \in F$ by lemma 4. f can be extended to a one-to-one mapping from P(x) onto $P(\omega_\alpha)$ by defining $f(y) = \{f(z); z \in y\}$ for $y \subseteq x$. But again $P(\omega_\alpha)$ is a wellfounded set. Since the aleph-hypothesis (AH) holds (in the set theory), there is a 1-1-mapping g from $P(\omega_\alpha)$ onto $\omega_{\alpha+1}$. Since both sets are in M, we obtain that $g \subseteq M$. But g is wellfounded, hence $H[g] = G \in F$ and therefore by (β) (see section B): g is in the model. Thus P(x) can be mapped one-to-one onto $\omega_{\alpha+1}$ and has therefore in $\mathfrak{M}$ power $\aleph_{\alpha+1}$, q.e.d.

Corollary. The statement (PW): "The powerset of a well-ordered set is wellorderable" holds in Fraenkel's-model, though (AC) fails in it.

Remark. H.Rubin has shown, that in full ZF the axiom of choice (AC) is equivalent to the statement (PW) (see: H.Rubin, Notices AMS, vol.7(1960)p.381, or H.+J.Rubin: Equivalents of the Ax.of choice, Amsterdam 1963,p.77-78). By our corollary (PW) $\rightarrow$ (AC) cannot be proved in ZF^0 alone: in ZF^0 the axiom of choice (AC) is independent from (PW), though both are equivalent in full ZF.

It follows from the result of H.Rubin, that in full ZF the (GCH) and the aleph-hypothesis (AH) are equivalent. Our lemma says that in ZF^0 alone the (GCH) is independent from (AH).

E) THE INDEPENDENCE OF THE AXIOM OF CHOICE (AC) FROM KUREPA'S ANTI-CHAIN PRINCIPLE

G.Kurepa has considered in his paper "Ueber das Auswahlaxiom", Math. Annalen 126(1953)p.381-384, the following statement

(KA) "Every partially ordered set $\langle s, \leqslant \rangle$ has a maximal antichain".

Here a subset t of s is called an antichain iff for $x,y \in t \rightarrow \neg(x \leqslant y \vee y \leqslant x)$. Kurepa has shown that in ZF^0, (KA) in conjunction with the

(O) Ordering-theorem: "Every set can be linearely ordered"

is equivalent to the axiom of choice: $ZF^0 \vdash (AC) \leftrightarrow [(KA) \wedge (O)]$ and has asked wether (KA) alone is equivalent to (AC) of not. We have proved (U.Felgner, Math.Zeitschr.111(1969)p.221-232) that in full ZF in fact (AC) and (KA) are equivalent (the axiom of foundation is used in the proof).In contrast to this result J.D.Halpern was able to show that in ZF^0 alone (AC) is independent from (KA). The proof is contained in Halpern's thesis (Berkeley 1962) and not yet published.

We shall work in a set theory ZF^0 + (AC) in which there is an infinite set A of reflexive sets such that $V = \bigcup_\alpha R_\alpha(A)$.

Let G be a group of permutations of A. If H is a subgroup of G consisting of those elements of G which leave a finite subset E of A pointwise fixed, then we say that H has finite support and E is called a support ("Träger") of H.

Definition of Halpern's model. Let G be the group of all permutations of $A = R_0(A)$ (a permutation is a surjective one-to-one mapping). Let F be the set of all such subgroups H of G which contain a finite-support subgroup. F is a filter of subgroups of G satisfying conditions (i),...,(iv). Define $\mathfrak{M} \triangleq \mathfrak{M}[G,F]$ as in section B. It follows that $\mathfrak{M}$ satisfies the axioms of ZF^0 and that $A = R_0(A)$ is a set of $\mathfrak{M}$ such that (WF), the weak axiom of foundation, holds in $\mathfrak{M}$ with respect to A as base.

Remark. Since G is the group of all permutations of A, the supports of the finite support-subgroups of G are uniquely determined (If G would not be the full group of permutations the supports would not be unique). If H is a finite support subgroup of G we let supp(H) be the support of H.

Lemma 1. If H_1, H_2 are finite support-subgroups of G, then $grp\{H_1,H_2\}$ is also a finite support-subgroup of G and $supp(grp\{H_1,H_1\}) = supp(H_1) \cap supp(H_2)$.

Proof. $grp\{H_1,H_2\}$ is the subgroup of G generated by H_1 and H_2, id est: the smallest subgroup of G containing H_1 and H_2. Define

$$H_3 = \{\tau \in G;\ \tau \text{ leaves } supp(H_1) \cap supp(H_2) \text{ pointwise fixed}\}$$

Clearly $grp\{H_1,H_2\} \leqslant H_3$. In order to prove the converse consider $\tau \in H_3$. Define $a = supp(H_1)$, $b = supp(H_2)$ and let

$$a - (a \cap b) = \{a_1,\dots,a_k\}$$

Let $c_1,\dots,c_k$ be distinct elements not in $a \cup \tau^{-1}(b) \cup b$. Let σ_1 be the permutation of $A = R_0(A)$ which permutes a_i with c_i $(1 \leq i \leq k)$ and is constant otherwise. Then $\sigma_1 \in H_2$. Let

$$b - (a \cap b) = \{b_1, \ldots, b_m\}$$

Notice that $\sigma_1\tau^{-1}(b_i) \notin a$ and $b_i \notin a$ (by definition) Let σ_2 be the permutation of A which maps b_i onto $\sigma_1\tau^{-1}(b_i)$ and is constant for the other elements of A. Hence

$$\sigma_2(b_i) = \sigma_1\tau^{-1}(b_i) \text{ and } \sigma_2 \in H_1 .$$

Finally define $\sigma_3 = \tau\sigma_1^{-1}\sigma_2$. Then

$$\sigma_3(b_i) = \tau\sigma_1^{-1}\sigma_2(b_i) = \tau\sigma_1^{-1}\sigma_1\tau^{-1}(b_i) = b_i$$

for $1 \leq i \leq m$. Further, since τ, σ_1 and σ_2 leave $a \cap b$ pointwise fixed, σ_3 also leaves $a \cap b$ pointwise fixed. Thus σ_3 leaves b pointwise fixed and we get $\sigma_3 \in H_2$. But $\sigma_3 = \tau\sigma_1^{-1}\sigma_2 \rightarrow \tau = \sigma_3\sigma_2^{-1}\sigma_1$, thus $\tau \in grp\{H_1, H_2\}$ q.e.d.

Lemma 2. If $H[x] = \{\tau \in G; \tau^*(x) = x\}$ includes a finite-support subgroup, then $H[x]$ contains a finite-support subgroup which includes all other finite-support subgroups which are contained in $H[x]$.

Proof. Remember that τ^* is the unique extension of τ to an automorphism of the universe V (see section B). Let I be the intersection over the set of supports of the finite-support subgroups included in $H[x]$. Since each support is finite, hence I is finite and can be represented as an intersection of only finitely many of supports. The group generated by the union of subgroups corresponding to these finitely many supports has support I (this follows from lemma 1 and an ordinary induction), is contained in $H[x]$ and is largest in the sense stated, q.e.d.

Remark and Definition. If x is in the model $\mathfrak{M}[G,F]$, then $H[x] \in F$. But by definition of the filter F, there is a finite support-subgroup H^* contained in $H[x]$. Thus, by lemma 2, $H[x]$ includes a largest finite-support subgroup. This subgroup is uniquely determined and depends only on $H[x]$ (if x is in $\mathfrak{M}[G,F]$), and we denote this finite-support subgroup of $H[x]$ by $H_0[x]$. Further we write $F(x) = supp(H_0[x])$.

We shall prove that Kurepa's Antichain Principle (KA) holds in $\mathfrak{M}[G,F]$. Since every "abstract" partial ordering $\leq$ on a set s can be represented by the inclusion relation $\subseteq$, we may restrict ourself to the discussion of sets t, where $\subseteq$ is the partial ordering on t (namely, for $x \in s$ define $[x] = \{y \in s;\ y \leq x\}$ and $t = \{[x];\ x \in s\}$, then $\langle s, \leq \rangle$ and $\langle t, \subseteq \rangle$ are isomorphic).

Lemma 3: (KA) holds in Halpern's model $\mathfrak{M}[G,F]$.

Proof. Let t be a set of $\mathfrak{M} \triangleq \mathfrak{M}[G,F]$, and define

$Z = \{y;\ y \subseteq t \wedge y \in \mathfrak{M} \wedge F(y) \subseteq F(t) \wedge y$ is an anti-chain$\}$.

Z is a subset of the model $\mathfrak{M}$ but not necessarily $Z \in \mathfrak{M}$. We want to prove that Z has maximal elements.

If T is a subset of Z, totally ordered by $\subseteq$, then $\bigcup T$ is again an antichain. Further $H[\bigcup T] \geq \bigcap \{H[y];\ y \in T\}$ and always: $y \in Z \to H_0[t] \leq H_0[y] \leq H[y]$ where $H_0[t] \in F$. Thus $H[\bigcup T] \in F$. Since $T \subseteq Z \subseteq \mathfrak{M}$, hence $\bigcup T \subseteq \mathfrak{M}$ by the transitivity of $\mathfrak{M}$ and therefore $\bigcup T \in \mathfrak{M}$ (see (α) and (β) in section B). Finally: $H_0[t] \leq H[\bigcup T] \to H_0[t] \leq H_0[\bigcup T]$, thus (by lemma 1): $F(\bigcup T) \subseteq F(t)$. This shows that $\bigcup T \in Z$. Since Z is inductively ordered by $\subseteq$ there is in Z by Zorn's lemma a maximal element, say y_0, in Z (notice that $Z \neq \emptyset$, since $F(\emptyset) = \emptyset \subseteq F(t)$, hence $\emptyset \in Z$). We want to show that y_0 is maximal among all antichains of $\langle t, \subseteq \rangle$ in $\mathfrak{M}$.

Suppose y_0 is not a maximal antichain of $\langle t, \subseteq \rangle$ in $\mathfrak{M}$. Then there is an element $y \in t - y_0$ such that $y_0 \cup \{y\}$ is an antichain.

Let $$y_1 = y_0 \cup \{\tau^*(y);\ \tau \in H_0[t]\}.$$

Then $H_0[t] \leq H[y_0]$ implies $H_0[t] \leq H[y_1]$. Since $H_0[t] \in F$ therefore $H[y_1] \in F$. Since $y_0 \subseteq y_1 \subseteq t \subseteq \mathfrak{M}$ therefore $y_1 \in \mathfrak{M}$ (by (β) in sect. B). On the other hand $H_0[t] \leq H[y_1]$ implies $F(y_1) \subseteq F(t)$. y_1 cannot be an antichain, since otherwise $y_1 \in Z$ contradicting the maximality of y_0. Hence y_1 must have two comparable elements. We have two cases:

Case 1: There are $z \in y_0$ and $\tau \in H_0[t]$ such that $z \subseteq \tau^*(y)$ or $\tau^*(y) \subseteq z$.

Case 2: There are $\tau_1, \tau_2 \in H_0[t]$ such that $\tau_2^*(y) \neq \tau^*(y)$ and $\tau_1^*(y) \subseteq \tau_2^*(y)$.

If case 1 holds, we have $(\tau^{-1})^*(z) \subseteq y$ or $y \subseteq (\tau^{-1})^*(z)$. But $\tau \in H_0[t] \leq H_0[y_0] \leq H[y_0]$ and $z \in y_0$ implies $(\tau^{-1})^*(z) \in y$. This contradicts the fact that $y_0 \cup \{y\}$ is an anti-chain.

If case 2 holds, put $\tau = \tau_1^{-1}\tau_2$. Since $H_0[t]$ is a group: $\tau \in H_0[t]$ and we obtain the existence of $\tau \in H_0[t]$ such that $y \neq \tau^*(y)$ and $y \subseteq \tau^*(y)$. We want to find a $m_0 \in \omega$ such that $\tau^{m_0 *}(y) = y$. Hence it is natural to look at:

$$D_1 = \{w \in F(y);\ \bigvee_n (0 < n \wedge \tau^n(w) = w)\}$$

$$D_2 = \{z \in A = R_0(A);\ \bigvee_w \bigvee_n [w \in D_1 \wedge \tau^n(w) = z]\}.$$

τ^n means n-times iterated application of τ. Since $F(y)$ is finite, D_1 is finite too. Further D_2 is finite since $w \in D_1$ implies that after a finite number of successive iterations of τ one comes back to w. Thus D_2 is finite since D_1 is finite and every element w of D_1 has only finitely many images under successive iteration of τ. Also, D_2 is closed under τ. Thus D_2 together with $\tau \upharpoonright D_2$ is a permutation group. Since τ is a one-to-one mapping on D_2, the finite cycles $S_z = \{z, \tau(z), \tau^2(z), \ldots\}$ are either equal of disjoint and form orbits of the permutation group $\langle D_2, \tau \upharpoonright D_2 \rangle = \mathfrak{G}$. The group $\mathfrak{G}$ is hence the direct sum of these cyclic groups. Let n_0 be the order of $\mathfrak{G}$.

If $z \in F(y) - D_1$, then $\tau^n(z)$ is never in D_2 (and there is a finite numver n_z such that $m > n_z$ implies $\tau^m(z) \notin F(y)$ (since $F(y)$ is finite and $\tau^n(z) \neq z$ for all $n > 0$). Let m_0 be the first multiple of n_0 strictly greater than $\mathrm{Max}\{n_z;\ z \in F(y) - D_1\}$. Then

$$\tau^{m_0}(w) = w \text{ if } w \in D_2$$

since $m_0 = k \cdot n_0$ for some $k \in \omega$ and the cardinality of the orbits of $\mathfrak{G}$ devide the order of $\mathfrak{G}$. Further by definition of m_0: $z \in F(y) - D_1 \rightarrow n_z < m_0$, hence $\tau^{m_0}(z) \notin F(y)$. Thus if $z \in F(y) \cap \{\tau^{m_0}(w);\ w \in F(y)\}$ then $\tau^{m_0}(z) = z$.

We define a permutation σ of $A = R_0(A)$ which maps $X = F(y)$ onto $Y = \{\tau^{m_0}(z);\ z \in F(y)\}$ by $\sigma(x) = \tau^{m_0}(x)$ for $x \in X$, $\sigma(u) = x$ for $u = \tau^{m_0}(x) \in Y$ and $\sigma(v) = v$ otherwise. σ is well-defined since on $X \cap Y$ we have that σ is the identity as was just proved above.

From $\sigma(z) = \tau^{m_0}(z)$ for $z \in F(y)$ we obtain

(1) $\sigma^*(y) = (\tau^{m_0})^*(y)$

since $\sigma \upharpoonright F(y) = \tau^{m_0} \upharpoonright F(y)$, hence $\sigma^{-1}\tau^{m_0} \in H_0[y] \leq H[y]$, thus $(\sigma^{-1}\tau^{m_0})^*(y) = y$, id est $\sigma^*(y) = \tau^{m_0 *}(y)$.

Since also $\sigma\tau^{m_0}$ is the identity on $F(y)$ we obtain in a quite similar way from $\sigma(\tau^{m_0}(z)) = z$ for $z \in F(y)$:

(2) $\sigma^*(\tau^{m_0})^*(y) = y$.

By hypothesis $y \subseteq \tau^*(y)$, and since τ^* is an automorphism, we have

(3) $y \subseteq \tau^*(y) \subseteq (\tau^2)^*(y) \subseteq \ldots \subseteq (\tau^{m_0})^*(y)$.

Thus by (1): $y \subseteq \sigma^*(y)$. But (3) also yields $\sigma^*(y) \subseteq \sigma^*(\tau^{m_0})^*(y)$. Applying (2), we have $\sigma^*(y) \subseteq y$. Thus $y = \sigma^*(y)$. From (1) we deduce $y = (\tau^{m_0})^*(y)$. Finally, from (3), we arrive at the contradiction $y = \tau^*(y)$ and **lemma** 3 is proved.

Let (AC_2) be the axiom of choice for families (= sets) whose elements are couples (= unordered pairs). Instead of proving that the (unrestricted) axiom of choice (AC) does not hold in Halpern's model $\mathfrak{M}$, we shall show that already (AC_2) fails in $\mathfrak{M}$.

Lemma 4: The weak axiom of choice (AC_2) does not hold in Halpern's model $\mathfrak{M}[G,F]$.

Proof: Let $Y = \{z; \bigvee_u \bigvee_v (u,v \in A = R_0(A) \wedge u \neq v \wedge z = \{\langle u,v \rangle, \langle v,u \rangle\})\}$. $A = R_0(A)$ is a set of $\mathfrak{M}$ as was noticed previously. Thus $Y \subseteq \mathfrak{M}$. But Y is closed under G, thus $H[Y] = G \in F$. Hence, by (β) of section B, Y is a set of the model $\mathfrak{M}$. Also $z \in Y$ implies $\overline{\overline{z}} = 2$ and distinct elements of Y are disjoint. Suppose there would be a choice set C for Y in $\mathfrak{M}$, id est $\bigwedge_w [w \in Y \rightarrow w \cap C$ has cardinality 1$]$ and $C \in \mathfrak{M}$. It follows $H[C] \in F$ and $F[C]$ is a finite subset of the infinite set $R_0(A)$. Pick elements $u,v \in R_0(A) - F(C)$ such that $u \neq v$. Let τ be the permutation of $R_0(A)$ which interchanges u and v and is the identity otherwise. Then $\tau \in H_0[C] \leqslant H[C]$, hence $\tau^*(C) = C$, and $y = \{\langle u,v \rangle, \langle v,u \rangle\} \in Y$. Suppose $\langle u,v \rangle \in C$, then $\tau^*(\langle u,v \rangle) = \langle v,u \rangle \in \tau^*(C)$, hence $\langle v,u \rangle \in C$: a contradiction. If $\langle v,u \rangle \in C$ then one concludes similarly that $\langle u,v \rangle \in C$, again contradicting the assumption on C. Thus Y has no choice set in $\mathfrak{M}$, q.e.d.

This finishes the proof, that in Halpern's model $\mathfrak{M}[G,F]$ all axioms of ZF^0, Kurepa's Antichain Principle (KA) and $\neg(AC_2)$ are true. As a

Corollary (J.D.HALPERN): The axiom of choice (AC) does not follow from Kurepa's Antichain-Principle (KA) in ZF^0.

Remark. Since (AC_2) fails in Halpern's model $\mathfrak{M}$, the ordering principle (0) fails in $\mathfrak{M}$ too, since $ZF^0 \vdash (0) \rightarrow (AC_2)$. Further $ZF^0 \vdash$ (BPI) $\rightarrow$ (0) (via compactness-theorem of the lower predicate calculus, e.g.) where (BPI) is the Boolean Prime Ideal theorem "Every Boolean algebra has a prime ideal". Stone (Trans.AMS vol.40(1936) p.37-111) has shown in ZF^0 that (BPI) is equivalent to the "Representation Theorem for Boolean Algebras": "Every Boolean Algebra $\langle B, \sqcup, \sqcap, \neg \rangle$ is isomorphic to a set-algebra $\langle C, \cup, \cap, - \rangle$".

The statement

(SPI): "Every infinite set algebra has a non-principle prime ideal"

follows from the (BPI). Tarski has asked, whether (SPI) → (BPI) is provable. Halpern has shown, that in ZF^0 this implication is not provable. Halpern shows that in the model above the (SPI) holds while the (BPI) fails in it.

Lemma 5 (U.Felgner, M.Z. 111(1969)): Kurepa's Antichain Principle (KA) implies in ZF^0 the statement (LW) which says that every lineareely ordered set can be well-ordered.

Proof. Let $\langle s, \leq \rangle$ be a linearly ordered set. The powerset P(s) of s is a set of chains. By a theorem of Zermelo (Math.Ann.65(1908)p. 261-281, theorem 28) there is a set K whose elements are pairwise disjoint, such that there is a one-to-one mapping f from P(s) - {∅} onto K with the property that for $\emptyset \neq t \in P(s)$, $f(t) \in K$ is isomorphic to t. Thus P(s) is represented isomorphically by K, but K is a set of pairwise disjoint chains. A maximal antichain C of $\bigcup K$ is a choice function which selects from each chain just one element. Thus we get a choice function g defined on P(s) - {∅}. By Zermelo's well-ordering theorem (Math.Ann.59(1904)p.514-516, or 65(1908)p.107-128) the set s can be wellordered, q.e.d.

Thus (LW) holds in Halpern's model $\mathfrak{M}$ too. Since $ZF^0 \vdash$ (LW) → (PW), we have strengthened our result in chapter D: in ZF^0 the (AC) is independent from (LW). Further, Felgner proved that (LW) → (KA) is not a theorem of ZF^0.

F) THE UNDEFINABILITY OF CARDINALITY IN ZF^0

One says that the sets x and y are equipotent (or equinumerous), in symbols $x \approx y$, iff there is a one-to-one function mapping x on y. The notion of the cardinal number $\bar{\bar{x}}$ of x is obtained from equipotency by abstraction. In the presence of the axiom of choice (AC) the term $\bar{\bar{x}}$ can be defined to be the least ordinal α equipotent with x. If we do not have (AC) but the axiom of foundation, we are still able to define adequately $\bar{\bar{x}}$ à la Frege-Russell-Scott:

$$\bar{\bar{x}} = \{y;\ y \approx x \wedge \bigwedge_z (z \approx x \rightarrow \rho(y) \leq \rho(z))\}$$

where ρ is the Mirimanoff-rank function (see chapt.I,sect.E). Here $\bar{\bar{x}}$ consists of sets y of lowest rank equinumerous with x (see D.Scott: Definitions by abstraction in axiomatic set theory, Bull.AMS 61(1955) p.442, and

[74] Dana SCOTT: The notion of rank in set-theory; Summaries Summer Institute for Symbolic Logic, Cornell Univ.1957, p.267-269).

We remark, that even in the absence of both the axioms of choice and regularity but in the presence of either the weak axiom of foundation in the form "there is a set A such that $V = \bigcup_\alpha R_\alpha(A)$" or the axiom (U.Fg.,Archiv d.Math.20): "the universe V can be covered by a well-ordered sequence of sets s_α, α an ordinal". We shall show that in ZF^0 without any additional covering axiom (like foundation, etc.) there is no adequate definition of the term $\bar{\bar{x}}$. This result was obtained first by Azriel Lévy

[50] A.LÉVY: The Definability of Cardinal Numbers; in: "Foundations of Mathematics", Gödel-Festschrift, Springer-Verlag Berlin 1969,p.15-38.

Also R.J.Gauntt has obtained this result (independently):

[22] R.J.GAUNTT: Undefinability of Cardinality; Proceedings of the U.C.L.A.-set Theory Institute 1967. To appear in 1970.

In the presentation of the proof we shall follow mainly R.J.Gauntt but in few details A.Lévy.

When one considers the question of whether one can define in ZF^0 the cardinality operation $\bar{\bar{x}}$, the following possibilities turn up:

(a) $\bar{\bar{x}}$ is definable in a set theory ST: there is a term t(x) of ST with the only free variable x such that

$$ST \vdash \bigwedge_x \bigwedge_y [t(x) = t(y) \leftrightarrow x \approx y]$$

(b) $\bar{\bar{x}}$ is relatively definable in a set theory ST: there is a term t(x, z) of ST with the only free variables z and x such that

$$ST \vdash \bigvee_z \bigwedge_x \bigwedge_y [t(x,z) = t(y,z) \leftrightarrow x \approx y].$$

Obviously (a) entails (b) (Lévy [50] considers further possibilities). If we take ZF^0 + foundation (id est ZF) or ZF^0 + (AC) as set theory ST, then (a) holds. We shall prove a strong undefinability result, namely,that even (b) does not hold for the set theory ZF^0.

Theorem (Lévy,Gauntt): If ZF^0 is consistent, then so is ZF^0 plus the schema

$$(*) \quad \neg \bigvee_x \bigwedge_a \bigvee_y [\Phi(y,a,x) \wedge \bigwedge_b (a \approx b \leftrightarrow \phi(y,b,x))].$$

Proof. If ZF^0 is consistent, then also the theory (called ZF^∇) ZF^0 + "there is a proper class A of reflexive sets equinumerous with On (the class of all ordinals), such that for every x there exists $y \in x$ with either $y \cap x = \emptyset$ or $y \in A$" is consistent (see the results of Chapt.III, sect.A). Hence there is a function G (a classterm) mapping On one-to-one onto A. We will now construct within this universe a Fraenkel-Mostowski-Specker model $\mathfrak{M}$ of ZF^0 plus the schema (•). - In the sequel the elements of A are called atoms.

Each ordinal α can be written (in a unique way) as $\beta + n$ where β is a limit ordinal and $n \in \omega$ (this follows from Cantor's normal-form theorem). Define $\alpha \equiv 0$ iff $n \equiv 0$ (congruence modulo 2) for $\alpha = \beta + n \wedge Lim(\beta) \wedge n \in \omega$, and define $\alpha \equiv 1$ iff $n \equiv 1$ modulo 2 for $\alpha = \beta + n \wedge Lim(\beta) \wedge n \in \omega$. The ordinals congruent o are thus $0,2,4,\ldots,\omega,\omega+2,\omega+4,\ldots$ and the ordinals congruent 1 are $1,3,5,\ldots,\omega+1,\omega+3,\ldots$ For each ordinal α, $\{G(\alpha),G(\alpha+1)\}$ is a pair of atoms and if $\alpha \equiv o$ then $\bigwedge_\beta [\beta \equiv o \wedge \alpha \neq \beta \rightarrow \{G(\alpha),G(\alpha+1)\} \cap \{G(\beta),G(\beta+1) = \emptyset]$.

Definition. $F(\alpha) = \{G(\beta); (\beta \equiv 0 \wedge \beta < \alpha) \vee (\beta \equiv 1 \wedge \beta \leq \alpha)\}$.

The following definition is due to D.Mirimanoff (L'Ens.Math.vol.17 (1917)p.33 and p.211).

Definition. $Ker(x) = C(x) \cap A$ = the set of atoms in the transitive closure of x.

(read: the kernel of x; Mirimanoff used the term "noyaux").
We now restrict the universe to elements of sets built up from $F(\alpha)$'s, i.e. $V = \bigcup_\alpha(\bigcup_\gamma R_\gamma(F(\alpha)))$. That is, the restricted universe consists of all x for which $\bigvee_\alpha \bigvee_y (x \in y \wedge Ker(y) \subseteq F(\alpha))$.

e.g. for $\alpha < \beta$:

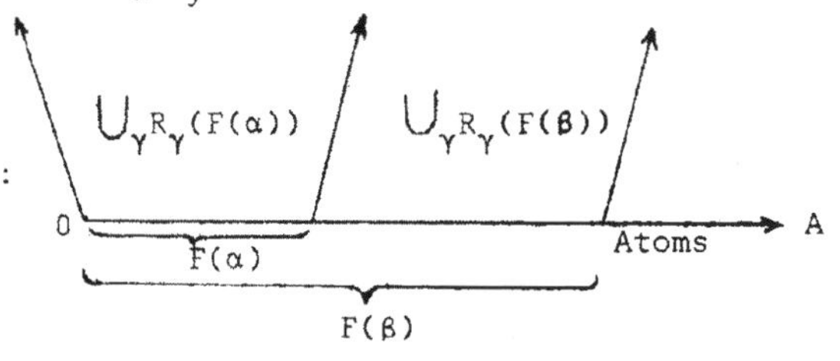

Notice that each $F(\alpha)$ is s set and $\bigcup_\alpha F(\alpha) = A$, where A is a proper class. For each permutation f on $F(\alpha)$, define f(x) over the entire (restricted) universe as follows:

$f(x) = x$ for atoms x not in $F(\alpha)$,

$f(x) = \{f(y);\ y \in x\}$ for sets x.

This definition is welldefined since if x is in the restricted universe, then $x \subseteq R_\gamma(F(\beta))$ for some ordinals β and γ [for the definition of $R_\gamma(a)$ see p.53]. By induction hypothesis it is assumed that f is defined for all $y \in R_\delta(F(\beta))$ for $\delta \leqslant \gamma$ and all β.

Definition. A permutation f on $F(\alpha)$ is called semi-admissible iff it preserves pairs, id est, for all $\beta < \alpha$ such that $\beta \equiv 0$ there exists $\gamma < \alpha$ such that $\gamma \equiv 0$ and

$$f(\{G(\beta),G(\beta+1)\}) = \{G(\gamma),G(\gamma+1)\}.$$

Definition. A permutation f on $F(\alpha)$ is called admissible iff it fixes pairs, id est: for all $\beta < \alpha$ such that $\beta \equiv 0$ it holds that

$$f(\{G(\beta),G(\beta+1)\}) = \{G(\beta),G(\beta+1)\}.$$

Definition. x is symmetric $\leftrightarrow$ there is a finite set $\underline{a}$ of atoms such that each admissible permutation τ which leaves $\underline{a}$ pointwise fixed, fixes x (not necessarily pointwise!).

Definition. Sets of the model $\mathfrak{M}$ are those sets x which are hereditarily symmetric (id est: x and every element y of the transitive closure of x is symmetric).

Digression. Notice the similarity of the model $\mathfrak{M}$ just defined with Fraenkel's model in section C. The admissible permutations of $F(\omega)$ are called there "nice". But there is one important difference. In the definition of a symmetric set x we avoided the use of the notion of a finite support subgroup $K[\underline{a}]$ and in the definition of the model $\mathfrak{M}$ we avoided the use of a filter of subgroups. This is done since the permutations are already proper classes. Hence the groups $K[\underline{a}]$ would be totalities of proper classes and the filter F a collection of those totalities. It is possible to formulate a type theoretic extension of ZF-set theory in which sets can be collected to classes (à la v.Neumann-Bernays-Gödel) and in which classes can be collected to totalities, totalities to systems etc (in which the predicates "set", "class", "totality" "system",... are primitiv) using an idea of I.L. Novak-Gàl (Fund. Math. 37(1951)p.87-110). In such a set theory one can talk about the groups $K[\underline{a}]$, the filter F, etc. But since in the discussion above reference is made only with respect to one single class of permutations f of the sets F(α) we could restrict ourself to mention only permutations of certain type. The use of the notions "subgroup", "filter" would make only linguistical differences. Further remark that a permutation of the class A of atoms moves only elements which are in some F(α). Hence the "essential" part of a permutation on A is a set. This explaines that in the definition of a symmetric set we have quantified only over sets (thus class-variables to range over permutations on A are not needed). The formulae: "x is symmetric" and "x is hereditarily symmetric" are thus ZF-formulae. Thus $\mathfrak{M} = \{x;\ x \text{ is hereditarily symmetric}\}$ is a class-term of ZF.

The following lemmata are easily proved. The proofs are similar to those of section B.

Lemma 1. (In ZF^∇): If f is semi-admissible and g is admissible, then $f^{-1}gf$ is admissible.

Lemma 2. (In ZF^∇): $x \in \mathfrak{M} \leftrightarrow (x \subseteq \mathfrak{M} \wedge x$ is symmetric).

Lemma 3. (In ZF^∇): No two disjoint infinite sets of atoms are equinumerous in $\mathfrak{M}$.

Proof. Suppose the lemma is false. Then there are such infinite sets x and y and a one-to-one function g, mapping x onto y, in $\mathfrak{M}$. Since x,y and g are symmetric, there are finite sets a,b,c of Atoms such that every admissible permutation π leaving a (resp. b,c) pointwise fixed, fixes x (resp. y,g). If $G(\alpha) \in x - a$ for $\alpha \equiv 0$, then $G(\alpha+1) \in x$. Now pick $\alpha \equiv 0$ such that $G(\alpha) \in x - (a \cup b \cup c)$. Since g maps x onto y, $g(G(\alpha)) = G(\beta) \in y$ and $g(G(\alpha+1)) = G(\gamma) \in y$. Thus $\langle G(\alpha),G(\beta)\rangle \in g$ and $\langle G(\alpha+1),G(\gamma)\rangle \in g$. Take an admissible permutation π which interchanges the atoms $G(\alpha)$ and $G(\alpha+1)$ but is the identity otherwise. Since x and y are disjoint, π acts as the identical mapping on y. Thus $\pi(\langle G(\alpha),G(\beta)\rangle) = \langle \pi(G(\alpha)),G(\beta)\rangle = \langle G(\alpha+1),G(\beta)\rangle \in \pi(g) = g$ since π leaves c pointwise fixed. Hence $g(G(\alpha+1)) = G(\beta)$, a contradiction, g would not be one-to-one, q.e.d.

Lemma 4. (In ZF^∇): Any permutation on F(α), which moves only finitely many atoms, is in $\mathfrak{M}$.

Proof. Let a be the finite set of atoms moved by the permutation π. Then every admissible permutation τ which leaves a pointwise fixed maps π onto itself.

Lemma 5. (In ZF^{∇}): For each x and semi-admissible permutation π,
$x \in \mathfrak{M} \leftrightarrow \pi(x) \in \mathfrak{M}$.

Proof. Use lemma 1 and lemma 2 and proceed as in the proof of lemma 1 of chapt. III, section B, page 55-56.

Lemma 6. For each ZF-formula $\Phi(x_1, \ldots, x_n)$ with n free variables, the following are theorems of ZF^{∇}:

(i) π semi-admissible $\rightarrow [\Phi(x_1, \ldots, x_n) \leftrightarrow \Phi(\pi(x_1), \ldots, \pi(x_n))]$

(ii) π semi-admissible $\rightarrow [\mathrm{Rel}(\mathfrak{M}, \Phi(x_1, \ldots, x_n)) \leftrightarrow \mathrm{Rel}(\mathfrak{M}, \Phi(\pi(x_1), \ldots, \pi(x_n)))]$.

Here $\mathrm{Rel}(\mathfrak{M}, \Phi)$ is the formula obtained from Φ by restricting all quantifiers to the class $\mathfrak{M}$ (see chapt. I, page 14). The proof is by induction on the length of Φ, using lemma 5.

Lemma 7. (In ZF^{∇}): $\mathfrak{M}$ is (with respect to ε) a model of ZF^{0}.

The proof is like the one of Specker's theorem (in section B, p.54) using lemmata 2, 5 and 6. Do not take the "hyper-classes" of all one-to-one mappings from A onto A, but take only the groups of admissible permutations on the sets $F(\alpha)$. These groups are sets! For every set x in the restricted universe only an initial segment $F(\alpha)$ of the class A of atoms is essential (definite).

Lemma 8. For each ZF-formula $\Phi(x_1, x_2, x_3)$ with three free variables, the following is provable in ZF^{∇}:

$$\mathrm{Rel}(\mathfrak{M}, \neg \bigvee_x \bigwedge_a \bigvee_y^1 [\Phi(y,a,x) \wedge \bigwedge_b (a \approx b \leftrightarrow \Phi(y,b,x))]).$$

Proof. Suppose that the lemma is false. Then there is a ZF-formula $\Phi(x_1, x_2, x_3)$ and a set x in $\mathfrak{M}$ as required above. Since x is in $\mathfrak{M}$, hence in the restricted universe, there is an ordinal α such that $\mathrm{Ker}(x) \subseteq F(\alpha)$, where $\alpha \equiv 0$ can be choosen. Define

$$D_1 = F(\alpha+\omega) - F(\alpha).$$

Cleary, $D_1 \in \mathfrak{M}$. Suppose y is the (unique) cardinal of D_1, where $y \in \mathfrak{M}$, id est $\mathrm{Rel}(\mathfrak{M}, \Phi(y, D_1, x))$.

Case 1. $\mathrm{Ker}(y) \subseteq F(\alpha)$.

Case 2. $\mathrm{Ker}(y) \not\subseteq F(\alpha)$.

If case 1 holds, define $D_2 = F(\alpha+\omega.2) - F(\alpha+\omega)$, where $\omega.2 = \omega+\omega$. Then $D_2 \in \mathfrak{M}$. There is a semi-admissible permutation π of the atoms:

$\pi(G(\alpha+n)) = G(\alpha+\omega+n)$

$\pi(G(\alpha+\omega+n)) = G(\alpha+n)$

$\pi(G(\beta)) = G(\beta)$ for $\beta < \alpha$ or $\alpha+\omega.2 \leqslant \beta$.

Thus π fixes each element of $F(\alpha)$ and takes D_1 onto D_2. Hence $\pi(D_1) = D_2$, $\pi(D_2) = D_1$, $\pi(x) = x$ and $\pi(y) = y$. Then

$$\mathrm{Rel}(\mathfrak{M},\Phi(y,D_1,x)) \leftrightarrow \mathrm{Rel}(\mathfrak{M},\Phi(\pi(y),\pi(D_1),\pi(x))) \leftrightarrow \mathrm{Rel}(\mathfrak{M},\Phi(y,D_2,x))$$

Hence y is also the cardinal of D_2. Thus $\mathrm{Rel}(\mathfrak{M}, D_1 \approx D_2)$ violating lemma 3.

If case 2 holds, there is an ordinal $\beta \geqslant \alpha$ such that $G(\beta) \in \mathrm{Ker}(y) \wedge G(\beta) \notin F(\alpha)$. Pick an ordinal $\gamma, \gamma \equiv \beta$, $\gamma > \beta$, and define a permutation τ on $D_1 \cup F(\gamma+1)$ which interchanges $G(\beta)$ and $G(\gamma)$ and interchanges $G(\beta+1)$ with $G(\gamma+1)$ iff $\beta \equiv 0$, and interchanges $G(\beta-1)$ with $G(\gamma-1)$ iff $\beta \equiv 1$. Since $\alpha \equiv 0$, hence $G(\delta) \in F(\alpha) \rightarrow \delta < \alpha$, τ fixes all elements of $F(\alpha)$. Thus $\tau(x) = x$. τ moves $\mathrm{Ker}(y)$ and hence $\tau(y) \neq y$. Clearly τ is semi-admissible. Hence by lemma 6:

$$\mathrm{Rel}(\mathfrak{M},\Phi(y,D_1,x)) \leftrightarrow \mathrm{Rel}(\mathfrak{M},\Phi(\tau(y),\tau(D_1),x)).$$

Thus $\tau(y)$ is the cardinal of $\tau(D_1)$. Since y is the cardinal of D_1 and $\tau(y) \neq y$, D_1 and $\tau(D_1)$ have different cardinality and are therefore not equinumerous in $\mathfrak{M}$. But by lemma 4, τ is a set of $\mathfrak{M}$ and is a one-to-one function in the sense of $\mathfrak{M}$. Thus D_1 and $\tau(D_1)$ would be equinumerous in $\mathfrak{M}$, a contradiction. Lemma 8 is thus proved.

The theorem of Lévy-Gauntt follows directly from lemmata 7 and 8.

G) A FINAL WORD

The main idea behind Gödel's construction of the model $\langle L,\varepsilon \rangle$ of ZF + (AC) was to make all sets of the model definable (or nameable) by means of a certain complexe language. The natural (inductively defined) wellordering of the language induced a wellordering of the model-class L. The main idea behind the construction of ZF^0-models $\mathfrak{M}$ in which choice fails is to guarantee that $\mathfrak{M}$ contains infinitely many sets of "indiscernible" sets. Then there is no reason why a function f defined on a infinite set of sets of mutually indiscernible elements should choose from each set just the one and not the other element. This was made precise by introducing the groups G of permutations on some infinite set $A = R_0(A)$ of "atoms" (reflexive sets) and the filter F of subgroups of G.

The symmetries of the model $\mathfrak{M}$ are determined by F. If x is in $\mathfrak{M}$ and. $x = \{\tau(y);\ \tau \in G\}$ for every $y \in x$, then x is a set of indiscernibles in $\mathfrak{M}$. In Fraenkel's model (see this chapter, section C) the sets $\{a_{2k}, a_{2k+1}\}$ are e.g. sets of indiscernibles. The set B of these sets of indiscernibles is the set-theoretical counterpart to Russell's sequence of pairs of (mutually indiscernible) socks.

The "classical" way for obtaining those families of sets of indiscernibles was to take an infinite sequence of "urelements" or "reflexive sets" and to take a certain nice permutation group which acts on them. The choice of the right permutation group and the right filter of subgroups is the alpha and omega in all applications of the Fraenkel-Mostowski-Specker method. The filter F defines on the group G a topology. If in the surrounding set theory the axiom of choice holds and the weak axiom of foundation such that the atoms form a set, then the corresponding model $\mathfrak{M}[G,F]$ satisfies the (AC) iff the topology is discrete (it is supposed that the filter F satisfies conditions (i),...,(iv)), and then the model coincides with the whole universe of sets. Thus, in order to get non-trivial applications of the FMS-method, the filter F has to contain never the trivial subgroup {1} of G.

In the next chapter we shall describe Cohen's forcing method. This method applies to full ZF-set theory and yields not only independence results "below" the (AC) but also the independence of V = L from the (GCH), the independence of (GCH) from (AC) and lots of further results. Again it is possible to introduce in Cohen-models indiscernible sets by destroying the (AC). We remark that it is even possible to construct ZF models in which V = L holds and which contain indiscernibles, but then one has to assume the existence of large cardinals κ satisfying the partition relation $\kappa \longrightarrow (\omega)_2^{<\omega}$, see J.Silver's paper: A large cardinal in the constructible universe, Fund.Math.69(1970)p.93-100.

Additions to chapter III

1) The part $K[x] \in F$ then there is a one-to-one mapping from x onto some well-founded set, of lemma 4 in section B, p.57-58, can be trivially proved as follows.

If $K[x] \in F$ then x is wellorderable in $\mathfrak{M}[G,F]$; namely let w be any wellordering of x, then $w \subseteq \mathfrak{M}[G,F]$. But obviously $K[x] \leq H[w]$, thus $w \in \mathfrak{M}[G,F]$. Thus x is wellorderable in $\mathfrak{M}$ and there are 1-1-mappings from x onto some ordinals in $\mathfrak{M}$. But ordinals are well-founded sets, Q.E.D.

2) The corollary on p.62 which says that (PW) holds in Fraenkel's model $\mathfrak{M}$ can be stregthened by asserting that even (LW) holds in $\mathfrak{M}$ while (AC) fails. Proof. Let $\langle s,\leq \rangle$ be a linearily ordered set in $\mathfrak{M}$. Define $R = \{\langle a,b \rangle ; a,b \in s \wedge a \leq b\}$; thus $H[R] \in F$ and $H[R] \leq H[s]$. We claim that for each $y \in s$ it holds that $H[R] \leq H[y]$. Suppose not, then there are $y \in s$ and a $\tau \in H[R]$ such that $\tau(y) \neq y$. But $\tau(y) \in s$ and R is a linear ordering on s, thus either $\langle y,\tau(y) \rangle \in R$ or $\langle \tau(y),y \rangle \in R$. If $\langle y,\tau(y) \rangle \in R$, then $\tau(\langle y,\tau(y) \rangle) = \langle \tau(y),\tau^2(y) \rangle = \langle \tau(y),y \rangle \in \tau(R) = R$, since $\tau^2 = 1$. But $\langle y,\tau(y) \rangle \in R \wedge \langle \tau(y),y \rangle \in R$ yields $y = \tau(y)$, a contradiction! The same argument applies to the case $\langle \tau(y),y \rangle \in R$. Thus every $\tau \in H[R]$ leaves s pointwise fixed. Thus, if w is any wellordering relation on s, then $H[R] \leq H[w]$ and it follows that $w \in \mathfrak{M}$, q.e.d.

3) It holds that $ZF^0 \vdash (AC) \rightarrow (LW) \rightarrow (PW)$, while $ZF \vdash (AC) \leftrightarrow (LW) \leftrightarrow (PW)$. We have shown under 2) that $(LW) \rightarrow (AC)$ is not provable in ZF^0. Using the model of Mostowski [64] one shows that $(PW) \rightarrow (LW)$ is not provable in ZF^0. Let us indicate that obviously Kinna-Wagners principle of choice of proper, non-empty subsets cannot hold in Mostowski's model, since (PW) holds in it and otherwise (AC) would be true in it (see Mostowski: Colloqu.Math.6(1958)p.207-208). Let us note further that J.D.Halpern has shown that in Mostowski's model the Boolean prime ideal theorem (BPI) holds (Fund.Math.55(1964) p.57-66.

4) Finally we refer to some important papers in which the FMS-method is applied: E.Mendelson [61], [62], and:

A.Mostowski: On the Principle of Dependent choices; Fund.Math.35 (1948)p.127-130: [68].

H.Läuchli: Auswahlaxiom in der Algebra; Comment.Math.Helvetica 37 (1962/63)p.1-18.

H.Läuchli: The Independence of the Ordering principle from a restricted axiom of choice; Fund.Math.54(1964)p.31-43.

CHAPTER IV

COHEN EXTENSIONS OF ZF-MODELS

In this chapter we study Cohen's forcing technique for constructing extensions of ZF-models. This technique was introduced in 1963 by Paul J.Cohen. Using this method Cohen has solved the long outstanding problems of the independence of the Continuum-hypothesis from the axiom of choice and the independence of the axiom of choice from the ZF-axioms (including foundation):

[9] P.J.COHEN: The Independence of the axiom of choice; mimeographed notes(32 pages), Stanford University 1963.

[10] P.J.COHEN: The Independence of the Continuum Hypothesis; Proc. Nat.Acad.Sci.USA, part 1 in vol.50(1963)p.1143-1148, part 2 in vol.51(1964)p.105-110.

A sketch of the proofs is contained in:

[11] P.J.COHEN: Independence results in set theory; In: The Theory of Models-Symposium, North Holland Publ.Comp.Amst.1965, p.39-54.

In these papers the constructible closure is obtained by means of Gödel's F(α)-hierarchy (Gödel's monograph [25], of 1940). Dana Scott has remarked that the constructible closure can be obtained in a much more elegant way using Gödel's M_α-hierarchy (Gödel's paper [24] of 1939). The presentation of the independence proofs in Cohen's monograph is based on these improvements:

[12] P.J.COHEN: Set Theory and the Continuum Hypothesis; New York - Amsterdam 1966 (Benjamin, Inc.).

Since the publication of Cohen's papers [9], [10] and [11] the forcing technique has been modified in various ways by several authors. Using modified "Gödel-functions F" W.Felscher and H.Schwarz have studied systematically Cohen-generic models (see Tagungsberichte Oberwolfach April 1965 and the dissertation of H.Schwarz: Ueber generische Modelle und ihre Anwendungen; Freiburg i.Br.1966). A topological approach to forcing has been developed by C.Ryll-Nardzewsky and G.Takeuti:

[83] G.TAKEUTI: Topological Space and forcing; Abstract in the J.S.L. vol.32(1967)p.568-569.

A detailed exposition of this approach is contained in:

[66] A.MOSTOWSKI: Constructible Sets with applications; Amsterdam - Warszawa 1969(North Holland + PWN).

That forcing can be understood as a boolean valuation of sentences has been discovered by D.Scott, R.M.Solovay and P.Vopěnka -see the forthcoming paper by Scott-Solovay, or Scott's lecture notes of the UCLA set theory Institute (August 1967) and :

[72] J.B.ROSSER: Simplified Independence Proofs; Academic Press 1969.

[86] P.VOPĚNKA: General theory of ∇-models; Comment.Math.Univ. Carolinae (Prague) vol.8(1967)p.145-170.

For further litterature on ∇-models see the bibliography in [86]. Some of Vopěnka's papers have been reviewed by K.Kunen in the J.S.L. 34(1969)p.515-516. -We shall present here the forcing method in a way close to P.J.Cohen, using ideas which are due to D.Scott, R.M.Solovay and others. The following basic publications will be useful:

[39] R.B.JENSEN: Modelle der Mengenlehre; Springer-Lecture Notes, vol.37, 1967.

[40] R.B.JENSEN: Concrete Models of Set Theory; In Sets, Models and Recursion theory, Leicester Proceedings 1965, North Holland PublComp.Amsterdam 1967, p.44-74.

[80] J.SILVER: Forcing à la Solovay; unpublished lecture notes (28 pages).

[51] A.LÉVY: Definability in axiomatic Set Theory I; in: Logic, Methodology and Philosophy of Sci., Congress Jerusalem 1964, North Holland Publ.Comp.Amst.1965, p.127-151.

The main difficulties which arise when one wants to extend a given ZF-model $\mathfrak{M}$ by adjoining some new sets $a_0, a_1, \ldots$ to $\mathfrak{M}$ are that the sets a_i may contain undesired information encoded by the interior $\in$-structure of a_i. For instance, the interior $\in$-structure of a_i may give rise to mappings which destroy the replacement axiom in the extension. These "new" sets a_i which, when added to $\mathfrak{M}$, <u>generate</u> a ZF-model are called "<u>generic sets</u>". The forcing method is a technique to obtain generic sets. Herein the main idea is that every finite part of the interior $\in$-structure of a_i has to be in $\mathfrak{M}$, id est, a_i has to fulfill finite amounts of conditions which can be posed in $\mathfrak{M}$. Then a determination of the whole interior $\in$-structure of a_i is obtained in a way similar to Lindenbaum's completing process (see e.g. Mendelson **[60]** p.64) by choosing a "complete sequence of conditions".

In this chapter we shall <u>not</u> construct socalled "endextensions" The extensions we are dealing with are those which contain the same ordinals!

A) THE FORCING RELATION IN A GENERAL SETTING

The simplest general framework for constructing Cohen models of ZF is provided by considering partially ordered structures. This approach, a straightforward generalization of Cohen's original work, is due to R.M.Solovay. We shall present here a slight generalization of Solovay's approach.

Let $\mathfrak{M} \triangleq \langle M, \in_M \rangle$ be a standard model of ZF (see p.25 for the definition of "standard"). Let

$$\mathfrak{A} = \langle A; R_i \rangle_{i \in I}$$

be a first-order relational system in $\mathfrak{M}$ with domain A and some n_i-ary relations R_i ($i \in I$) defined on A. We assume that A is a set in the sense of $\mathfrak{M}$. We want to extend $\mathfrak{M}$ by adding to $\mathfrak{M}$ a generic copy of $\mathfrak{A}$. The properties which this copy has to fulfill in the extension $\mathfrak{N}$ of $\mathfrak{M}$ are expressed in a certain formal language $\mathcal{L}$. The language $\mathcal{L}$ describes $\mathfrak{N}$. Since $\mathcal{L}$ shall express in a very detailed way all that what "happens" in $\mathfrak{N}$, we construct $\mathcal{L}$ as a ramified language which has means to talk about every v.Neumann-Stufe V_α separately. Formally this is done by introducing limited comprehension terms E^α (intended interpretation of $E^\alpha x\Phi(x)$: set of sets x of rank less than α satisfying Φ; the E is taken from the french word "Ensemble") and limited quantifiers $\bigvee^\alpha$ (read $\bigvee^\alpha_x \Phi(x)$ as: "there exists an x of rank less than α such that $\Phi(x)$).

<u>The Alphabeth of the ramified language $\mathcal{L}$</u>

1) One sort of set-variables: $v_0, v_1, v_2, \ldots, v_n, \ldots$ ($n \in \omega$). x,y,z,.. are used to stand for these variables.
2) Set-constants $\underline{x}$ for each set x of $\mathfrak{M}$.

3) Constants $\mathring{a}_j$ for each $j \in A$.
4) n_i-ary predicates π_i for each $i \in I$ and ε for membership.
5) logical symbols: $\neg$, $\vee$, $\bigvee$ (not, or, there exists).
6) limited comprehension operatirs E^α and limited quantifiers $\bigvee^\alpha$ for each ordinal α of $\mathfrak{M}$, and finally brackets.

It is possible to arrange that these symbols are sets of $\mathfrak{M}$ in the following way: $\neg = \langle 0,0 \rangle$, $\vee = \langle 0,1 \rangle$, $\bigvee = \langle 0,2 \rangle$, $\bigvee^\alpha = \langle 1,\alpha \rangle$, $\varepsilon = \langle 0,3 \rangle$, $v_i = \langle 0,4+i \rangle$, $E^\alpha = \langle 2,\alpha \rangle$, $\underline{x} = \langle 3,x \rangle$, $\mathring{a}_j = \langle 4,j \rangle$, $\pi_i = \langle 5,i \rangle$ and $(= \langle 6,0 \rangle$, $) = \langle 6,1 \rangle$.
The formulae of $\mathcal{L}$ are obtained from these symbols by concatenation as usual by recursion. It follows that the collection of all formulae constitutes a class of $\mathfrak{M}$.

Definition. The notions of a ranked (= limited) formulae and of a limited comprehension term of $\mathcal{L}$ are defined as follows:

(a) If $u_1, u_2, \ldots$ are limited comprehension terms, set-constants or constants $\mathring{a}_j$ or variables, then $u_1 \varepsilon u_2$ and $\pi_i(u_1,\ldots,u_{n_i})$ are limited formulae.

(b) If Φ and Ψ are limited formulas, then so are $\neg\Phi$, $\Phi \vee \Psi$ and $\bigvee_x^\alpha \Phi$ (for α in $\mathfrak{M}$).

(c) If Φ is a limited formula with no free variables other than x, and α is an ordinal of $\mathfrak{M}$ such that (i) Φ contains no occurrence of $\bigvee^\beta$ with $\beta > \alpha$, (ii) Φ contains no occurrence of E^β with $\beta \geqslant \alpha$, (iii) Φ contains no set-constant $\underline{x}$ for a set x of Mirimanoff-rank $\geqslant \alpha$, (iv) if $\alpha \leqslant \lambda$ then Φ contains no occurrence of $\mathring{a}_j$, then $E^\alpha x\Phi(x)$ is limited comprehension term.

The notion of a free variable is defined as usual; a limited formula without free variables is said to be a limited sentence. We shall refer to the set-constants, constants of the form $\mathring{a}_j$ ($j \in A$) and the limited comprehension terms as <u>constant terms</u>. Remark that the definition above of a limited comprehension term is given with respect to the parameter $\lambda \geqslant \omega$. In most applications we choose λ to be ω respectively $\omega+1$.

Definition. Let $\rho(x)$ be the Mirimanoff-rank of the set x in the sense of $\mathfrak{M}$ (see p.14). The degree $\delta(t)$ of a constant term t is given by:

(a) $\delta(\underline{x}) = \rho(x)$,
(b) $\delta(\mathring{a}_j) = \lambda$
(c) $\delta(E^\alpha x\Phi(x)) = \alpha$

Abbreviations. Let u and v be constant terms or variables; then $u = v$ stands for $\bigwedge_x (x \varepsilon u \leftrightarrow x \varepsilon v)$ where x is a variable distinct from u,v. For constant terms u and v, $u \simeq v$ will stand for $\bigwedge_x^{\alpha} (x \varepsilon u \leftrightarrow x \varepsilon v)$ where $\alpha = \mathrm{Max}\{\delta(u),\delta(v)\}$. $u \simeq v$ is thus a limited sentence.

Next we define in $\mathfrak{M}$ a well-founded, localizable partial-ordering between limited formulas Φ by assigning to Φ an ordinal $\mathrm{Ord}(\Phi)$ of $\mathfrak{M}$. Read $\mathrm{Ord}(\Phi)$ as "the order of Φ". This then allows to define in $\mathfrak{M}$ the the forcing relation $\Vdash$ between "conditions" and limited formulas by induction on $\mathrm{Ord}(\Phi)$. Obviously instead of defining $\mathrm{Ord}(\Phi)$ to be the ordinal $\omega^2.\alpha + \omega.e + m$ we could define $\mathrm{Ord}(\Phi)$ to be $\langle \alpha, e, m \rangle$ and then taking the lexicographical ordering to these triples.

Definition. For a limited formula Φ define

$$\mathrm{Ord}(\Phi) = \omega^2.\alpha + \omega.e + m$$

where (i) α is the least ordinal such that Φ contains no quantifier $\bigvee^{\beta}$ with $\beta > \alpha$ and no constant term t of degree $\geqslant \alpha$,

(ii) $e = 3$ iff Φ contains at least one of the symbols π_i, $e = 2$ iff Φ does not contain any π_i but Φ contains at least one of the symbols $\mathring{a}_j$, $e = 1$ iff Φ contains no symbol π_i and no symbol $\mathring{a}_j$ but Φ contains a subformula $v \varepsilon u$ where v is either a constant term with $\delta(v) + 1 = \alpha$ or a variable which stands in the scope of a limited quantifier $\bigwedge^{\alpha}$ (for α defined in (i)), $e = 0$ in all other cases.

(iii) m is the length of Φ.

Let S be an infinite set in $\mathfrak{M}$ such that $x \varepsilon S \rightarrow \rho(x) < \lambda$ and $\lambda = \sup\{\rho(x); x \varepsilon S\}$ where $\lambda \geqslant \omega$. We want to find for each $j \in A$ (where $\mathfrak{A} = \langle A, R_i \rangle_{i \varepsilon I}$ is the given first order relational system) a generic subset a_j of S and generic relations B_i for $i \in I$ between these a_j's such that in the metatheory $\mathfrak{A}$ and $\langle \{a_j; j \in A\}, B_i \rangle_{i \in I}$ are isomorphic. Id est: we want to find a generic copy of $\mathfrak{A}$.
The sets a_j have to fulfill certain properties, or in different words: they have to satisfy certain conditions p (like "$7 \varepsilon a_j$" for instance, or others) which can be posed in $\mathfrak{M}$.
Instead of defining the sets a_j directly we first give a list saying that the sets a_j and the relation B_i have in $\mathfrak{N}$ (the extension of $\mathfrak{M}$) only those properties which are "forced" by some finite amount of information.

<u>Definition</u>: A condition p is a finite partial function from $S \times A$ into $2 = \{0,1\}$.

Let P be the set of $\mathfrak{M}$ of all conditions and let $\leq$ be the partial ordering in P defined by $p \leq q \leftrightarrow p \subseteq q$.
The definition of forcing is given first for limited sentences Φ by induction on $Ord(\Phi)$. Notice that p varies over the <u>set</u> P and that for a given ordinal β, all the ranked sentences Φ with $Ord(\Phi) < \beta$ constitute a set in $\mathfrak{M}$. Therefore (by the recursion theorem) the definition of $p \Vdash \Phi$ by induction on $Ord(\Phi)$ is permissible.

<u>Definition</u> of the (strong) forcing relation $\Vdash$ for limited sentences Φ. The definition takes place in $\mathfrak{M}$. Let T be the $\mathfrak{M}$-class of constant terms and let u be a variable ranging over constant terms.

(1) $p \Vdash u \varepsilon \underline{x} \leftrightarrow (\exists y \in x)(p \Vdash u \simeq \underline{y})$.

(2) $p \Vdash u \varepsilon E^{\alpha}x\Phi(x) \leftrightarrow (\exists t \in T)(\delta(t) < \alpha \;\&\; p \Vdash u \simeq t \;\&$ $p \Vdash \Phi(t))$.

(3) $p \Vdash u \varepsilon \dot{a}_j \leftrightarrow (\exists x \in S)(p \Vdash \underline{x} \simeq u \;\&\; p(\langle x,j\rangle) = 1)$.

(4) $p \Vdash \neg\Phi \leftrightarrow \sim(\exists q \geq p)(q \Vdash \Phi)$.

(5) $p \Vdash \Phi \vee \Psi \leftrightarrow (p \Vdash \Phi \;\dot{\vee}\; p \Vdash \Psi)$.

(6) $p \Vdash \bigvee_x^{\alpha}\Phi(x) \leftrightarrow (\exists u \in T)(\delta(u) < \alpha \;\&\; p \Vdash \Phi(u))$.

(7) $p \Vdash \pi_i(u_1,\ldots,u_{n_i}) \leftrightarrow (\exists j_1,\ldots,j_{n_i} \in A)(\langle j_1,\ldots,j_{n_i}\rangle \in R_i \;\&$ $\&\; p \Vdash u_1 \simeq \dot{a}_{j_1} \;\&\; \ldots \;\&\; p \Vdash u_{n_i} \simeq \dot{a}_{j_{n_i}})$.

To see that $p \Vdash \Phi$ is indeed defined by induction on $Ord(\Phi)$, notice that the formulae occurring on the right side of $\leftrightarrow$ have order strictly smaller that the formulae occurring on the left side of $\leftrightarrow$. Further remark that in the definition of $Ord(u_1 \simeq u_2)$ we have $e = 1$.

The definition of $p \Vdash \Phi$ for arbitrary $\mathcal{L}$-sentences Φ will be given in the Metalanguage (and not in $\mathfrak{M}$) by induction on the (ordinary) length of Φ. This definition will be valid since p ranges over a set P and the collection of all formulae of $\mathcal{L}$ constitutes a set in the sense of the meta-theory (since $\mathfrak{M}$ is a set in the sense of the meta-theory). Again let u,v range over T and p,q range over P.

Definition of $p \Vdash \Phi$ for arbitrary (unlimited) $\mathcal{L}$-sentences Φ.

(8) $p \Vdash u \varepsilon v$ and $p \Vdash \pi_i(u_1,..,u_{n_i})$ are defined as above.

(9) $p \Vdash \neg \Phi \Leftrightarrow \sim(\exists q \geq p)(q \Vdash \Phi)$.

(10) $p \Vdash \Phi \vee \Psi \Leftrightarrow (p \Vdash \Phi \dot{\vee} p \Vdash \Psi)$.

(11) $p \Vdash \bigvee_x^{\alpha} \Phi(x) \Leftrightarrow (\exists u \in T)(\delta(u) < \alpha \ \& \ p \Vdash \Phi(u))$.

(12) $p \Vdash \bigvee_x \Phi(x) \Leftrightarrow (\exists u \in T)(p \Vdash \Phi(u))$.

It is obvious that for limited sentences Φ of $\mathcal{L}$, $p \Vdash \Phi$ according to this definition iff $p \Vdash \Phi$ according to the former definition. The rest of this section is devoted to the study of the formal properties of the forcing relation $\Vdash$. In the following three lemmata let Φ be any $\mathcal{L}$-sentence.

Consistency-Lemma. For no $p \in P$ do we have both $p \Vdash \Phi$ and $p \Vdash \neg \Phi$.

Proof. If $p \Vdash \Phi$ and $p \Vdash \neg \Phi$ for some $p \in P$ and some $\mathcal{L}$-formula Φ, then by (9) $p \Vdash \neg \Phi \rightarrow \sim p \Vdash \Phi$ and we get a contradiction in the metalanguage, q.e.d.

First Extension Lemma. If $p \Vdash \Phi$ and $p \leq q$, then $q \Vdash \Phi$.

Proof by induction on the complexity of Φ (i.e. for limited sentences Φ by induction on Ord(Φ) and for unlimited Φ by induction on the length of Φ), see e.g. Jensen [39] p.94-95.

Second Extension Lemma. For every $p \in P$ there is a $q \in P$, $p \leq q$, such that either $q \Vdash \Phi$ or $q \Vdash \neg \Phi$.

Proof. Suppose that for no $q \geq p$ we do have $q \Vdash \Phi$. Then $p \Vdash \neg \Phi$ by (9). Suppose now that for no $q \geq p$ we do have $q \Vdash \neg \Phi$. Then by (9): $p \Vdash \neg(\neg \Phi)$. But applying (9) twice one gets

$$p \Vdash \neg\neg \Phi \Leftrightarrow \sim(\exists q \geq p)[\sim(\exists q' \geq q)(q' \Vdash \Phi)]$$
$$\Leftrightarrow (\forall q \geq p)(\exists q' \geq q)(q' \Vdash \Phi)$$

Thus there exists $q' \geq p$ such that $q' \Vdash \Phi$, q.e.d.

Remark that forcing does not obey some simple rules of the propositional calculus. Exempla gratia, p may force $\neg\neg\Phi$ but not Φ. Furthermore, the forcing relation $\Vdash$ has by definition (clauses (5), (10), (12)) a homomorphism property with respect to disjunction $(\vee, \dot{\vee})$ and existential quantification $(\bigvee, \exists)$. If we introduce conjunction $\wedge$ and universal quantification $\bigwedge$ as usual, then one notices that $\Vdash$ does not have the homomorphism property for conjunction $(\wedge, \&)$ or for universal quantification $(\bigwedge, \forall)$. For example only

$$p \Vdash \Phi \wedge \Psi \Rightarrow (\exists q_1 \geq p)(\exists q_2 \geq p)[q_1 \Vdash \Phi \ \& \ q_2 \Vdash \Psi]$$

holds. We shall introduce a relation $\Vdash^*$ (called weak forcing), which has the property that $p \Vdash^* \Phi \Leftrightarrow p \Vdash^* \neg\neg\Phi$ and the homomorphism property for conjunction and universal quantification. $\Vdash^*$ does not have the homomorphism property for disjunction and existential quantification and is, as we may say, dual to the strong forcing relation $\Vdash$.

Definition. $p \Vdash^* \Phi \Leftrightarrow p \Vdash \neg(\neg \Phi)$ "p weakly forces Φ"

$p \parallel \Phi \Leftrightarrow (p \Vdash \Phi \ \dot{\vee}\ p \Vdash \neg \Phi)$ "p decides Φ"

$p \parallel^* \Phi \Leftrightarrow (p \Vdash^* \Phi \ \dot{\vee}\ p \Vdash^* \neg \Phi)$ "p weakly decides Φ"

$\Vdash \Phi \Leftrightarrow (\forall p \in P)(p \Vdash \Phi)$.

p_1 and p_2 are compatible $\Leftrightarrow (\exists q \in P)(p_1 \leq q \ \& \ p_2 \leq q)$.

Lemma A: The weak forcing relation has the following properties (u, v are variables for terms and Φ, Ψ are any $\mathcal{L}$-formulae)

(i) $p \Vdash^* \Phi \Leftrightarrow \sim(\exists q)[p \leq q \ \& \ q \Vdash \neg \Phi]$,

(ii) $p \Vdash \Phi \Rightarrow p \Vdash^* \Phi$,

(iii) $p \Vdash^* \neg \Phi \Leftrightarrow p \Vdash \neg \Phi$,

(iv) If Φ is of the form $\Psi_1 \wedge \Psi_2, \Psi_1 \Leftrightarrow \Psi_2, \bigwedge_x \Psi, \bigwedge_x^\alpha \Psi$, $u = v$ or $u \approx v$, then $p \Vdash \Phi \Leftrightarrow p \Vdash^* \Phi$,

(v) $p \Vdash^* \Phi \wedge \Psi \Leftrightarrow [p \Vdash^* \Phi \ \& \ p \Vdash^* \Psi]$,

(vi) $p \Vdash^* \bigwedge_x \Phi \Leftrightarrow (\forall u \in T)[p \Vdash^* \Phi(u)]$,

(vii) $p \Vdash^* \bigwedge_x^\alpha \Phi \Leftrightarrow (\forall u \in T)[\delta(u) < \alpha \Rightarrow p \Vdash^* \Phi(u)]$,

(viii) $p \Vdash^* \Phi \leftrightarrow \Psi \Rightarrow [p \Vdash^* \Phi \Leftrightarrow p \Vdash^* \Psi]$,

(ix) $(\forall p' \geq p)(\exists q \geq p')[q \Vdash^* \Phi \Leftrightarrow q \Vdash^* \Psi] \Rightarrow p \Vdash^* \Phi \leftrightarrow \Psi$,

(x) $(\forall q \geq p)[q \Vdash^* \Phi \Leftrightarrow q \Vdash^* \Psi] \Rightarrow p \Vdash \Phi \leftrightarrow \Psi$.

Proof. Ad(i): Let Ψ be $\neg\Phi$. By (9) of the forcing definition $\sim(\exists q \geq p)[q \Vdash \Psi]$ is equivalent to $p \Vdash \neg\Psi$ which is $p \Vdash \neg\neg\Phi$; by definition of $\Vdash^*$ this is equivalent to $p \Vdash^* \Phi$.

Ad(ii): follows from the first extension lemma and (9).

Ad(iii): "$\Leftarrow$" follows from (ii). Now assume $p \Vdash^* \neg\Phi$ and suppose that $\sim p \Vdash \neg\Phi$. Then by (9): $q \Vdash \Phi$ for some $q \geq p$. Thus by (ii) $q \Vdash^* \Phi$. This is in contradiction with $p \Vdash^* \neg\Phi$ by the consistency lemma.

Ad(iv): Notice that all the forms of Φ listed are of the form $\neg\Gamma$, thus the claim follows from (iii). The symbols $\wedge, \leftrightarrow, \bigwedge, \bigwedge^\alpha$ are introduced by definition for longer expressions in terms of $\neg, \vee, \bigvee, \bigvee^\alpha$ only.

Ad(v): $p \Vdash^* \Phi \wedge \Psi$ is by (iv) equivalent with $p \Vdash \Phi \wedge \Psi$, which is by definition: $p \Vdash \neg(\neg\Phi \vee \neg\Psi)$. This is, using first (9) and then (10) of the forcing definition equivalent to

$$(\forall q)[p \leq q \Rightarrow \sim(q \Vdash \neg\Phi \dot{\vee} q \Vdash \neg\Psi)].$$

Using again (9) one gets equivalently $p \Vdash \neg\neg\Phi$ & $p \Vdash \neg\neg\Psi$.

Ad(vi): The proof is similar to the proof of (v)

Ad(vii): Again the proof similar to (v) or (vi).

Ad(viii): Assume $p \Vdash^* \Phi \leftrightarrow \Psi$ and $p \Vdash^* \Phi$ but $\sim p \Vdash^* \Psi$. By (i) there is an extension q of p such that $q \Vdash \neg\Psi$. Since $p \leq q$ the first extension lemma yields $q \Vdash \neg\neg\Phi$. Thus by (9) of the forcing definition

$$(\forall q')[q \leq q' \Rightarrow \sim(q' \Vdash \neg\Phi \dot{\vee} q' \Vdash \Psi)].$$

Using (10) and then again (9) of the forcing definition this gives us $q \Vdash \neg(\neg\Phi \vee \Psi)$. Thus: $q \Vdash \neg(\Phi \rightarrow \Psi)$ by definition of $\rightarrow$. Using (iii) and (v) one sees that this is in contradiction with $p \Vdash^* \Phi \leftrightarrow \Psi$.

Ad(ix): suppose that the conclusion does not hold and proceed using (i) and (9), (10) of the forcing definition and the second extension lemma. In this way one immediately gets a contradiction.

Ad(x): follows directly from (ix). This proves lemma A.

Lemma B. Let p and q be elements of the set P of conditions and let Φ and Ψ be $\mathcal{L}$-sentences.

(i) If p_1 and p_2 are compatible and $p_1 \Vdash \Phi$ and $p_2 \Vdash \Psi$, then $q \Vdash \Phi \wedge \Psi$ for every q greater than both, p_1 and p_2.

(ii) If $p \parallel \Phi$ and $p \parallel \Psi$ then $p \parallel \Phi \vee \Psi$, $p \parallel \Phi \rightarrow \Psi$, $p \parallel \Phi \wedge \Psi$ and $p \parallel \Phi \leftrightarrow \Psi$.

(iii) If $p \parallel \Phi$ and $p \parallel \Psi$, then $p \Vdash \Phi \wedge \Psi \Leftrightarrow (p \Vdash \Phi \;\&\; p \Vdash \Psi)$.

(iv) If $p \parallel \Phi$ and $p \parallel \Psi$, then $p \Vdash \Phi \leftrightarrow \Psi \Leftrightarrow (p \Vdash \Phi \Leftrightarrow p \Vdash \Psi)$.

Proof by direct computation.

Lemma C. Let p, Φ and Ψ be as in lemma B.

(i) If $p \parallel^* \Phi$ and $p \parallel^* \Psi$, then $p \parallel^* \neg \Phi$, $p \parallel^* \Phi \wedge \Psi$, $p \parallel^* \Phi \vee \Psi$, $p \parallel^* \Phi \rightarrow \Psi$ and $p \parallel^* \Phi \leftrightarrow \Psi$.

(ii) If $p \parallel^* \Phi$ and $p \parallel^* \Psi$, then $p \parallel^* \Phi \vee \Psi \Leftrightarrow (p \Vdash^* \Phi \;\dot\vee\; p \Vdash^* \Psi)$.

(iii) If $p \parallel^* \Phi$ and $p \parallel^* \Psi$, then $p \Vdash^* \Phi \leftrightarrow \Psi \Leftrightarrow (p \Vdash^* \Phi \Leftrightarrow p \Vdash^* \Psi)$.

Proof by direct computation (use lemma A, (viii) and (ix)).

Lemma D. If $p \parallel \Phi_i$ $(i = 1,\dots,n)$, C is an n-ary sentential connective id est: an operation which is an iteration of the primitive sentential connectives $\neg$ and $\vee$) and $\dot{C}$ is the corresponding sentential connective of the meta-language (id est: the analog of $\sim$ and $\dot\vee$), then:

(i) $p \parallel C(\Phi_1,\dots,\Phi_n)$, and

(ii) $p \Vdash C(\Phi_1,\dots,\Phi_n) \Leftrightarrow \dot{C}(p \Vdash \Phi_1,\dots,p \Vdash \Phi_n)$.

Proof by induction on the number of times $\neg$ and $\vee$ are used in C (use lemma B).

Lemma E. Let C be a n-ary sentential connective. If $C(\phi_1,\dots,\phi_n)$ is a tautology for all $\phi_1,\dots,\phi_n$, then for all p and for all $\mathcal{L}$-sentences $\Phi_1,\dots,\Phi_n$ it holds that

$$p \Vdash^* C(\Phi_1,\dots,\Phi_n).$$

Proof. Use lemma A and C (see e.g. A.Lévy [51] p.141).

Lemma F. Let u, v and w stand for constant terms; then for every p:

(i) $p \Vdash u = u$,

(ii) $p \Vdash u = v \Rightarrow p \Vdash v = u$,

(iii) $[p \Vdash u = v \;\&\; p \Vdash v = w] \Rightarrow p \Vdash u = w$,

Lemma G. Again let u, v and w be constant terms, then for every p:

(i) $p \Vdash u \simeq v \Leftrightarrow p \Vdash u = v$,

(ii) $[p \Vdash u \in w \;\&\; p \Vdash u = v] \Rightarrow p \Vdash v \in w$,

(iii) $[p \Vdash w \in u \;\&\; p \Vdash u = v] \Rightarrow p \Vdash^* w \in v$.

For a proof of lemmata F and G see Lévy [51] p.141 or Easton, Annals of Math. Logic, vol.1(1970):[14].

Corollary H: If $p \Vdash \pi_i(u_1,\ldots,u_{n_i})$ and $p \Vdash u_1 = v_1, \ldots p \Vdash u_{n_i} = v_{n_i}$, then $p \Vdash \pi_i(v_1,\ldots,v_{n_i})$.

This follows easily from the definition of forcing, clause (7), and lemmata F and G.

Digression: The forcing definition $p \Vdash \Phi$ between elements p of the set of conditions (of $\mathfrak{M}$) and limited $\mathcal{L}$-formulae Φ was given in the "groundmodel" $\mathfrak{M}$ while the definition of $p \Vdash \Phi$ for unlimited Φ was given in the underlying meta-theory. We shall show in the sequel that for each specific $\mathcal{L}$-sentence Φ the forcing relation can be defined in $\mathfrak{M}$, because Φ is finite and the construction of the class K_Φ of p's forcing Φ can be done in finitely many steps. For each specific Φ the mechanism of constructing K_Φ can be implemented within $\mathfrak{M}$ but the mechanism is not universally applicable for all sentences Φ of $\mathcal{L}$, so that within $\mathfrak{M}$ we do not have the whole relation $\Vdash$. This is not too much surprising, since the definition of forcing resembles very much the definition of truth, and by the Epimenides-Tarski paradox we cannot define in ZF (or within the $\mathfrak{M}$-language) the notion of truth for $\mathcal{L}$-sentences [see Lévy [51] p.138, A.Tarski: Logic, Semantics, Metamathematics (Oxford 1956) p. 248, Fraenkel-BarHillel: Foundation of Set Theory (Amsterdam 1958) p.306 and Kleene: Introduction to Meta-Mathematics (Amsterdam-Groningen 1967) p.39,42, 501, see also Mendelson [60] p.151]. However we can define forcing for a single given sentence Φ or for some particular family of sentences within $\mathfrak{M}$.

Lemma I: Let $\Phi(x_1,\ldots,x_n)$ be an unlimited formula of $\mathcal{L}$. There is a class K_Φ of the model $\mathfrak{M}$ whose elements are the (n+1)-tuples $\langle p,u_1,\ldots,u_n\rangle$ such that $p \Vdash \Phi(u_1,\ldots,u_n)$, where the $u_i (1 \leq i \leq n)$ are constant terms.

According to our remark on page 79 the constant terms u_i are considered as certain special finite sequences of symbols which are in $\mathfrak{M}$ - for more details see Easton's thesis, Annals of math.Logic, vol 1(1970).

Proof by induction on the length of the formula Φ. Since the atomic formulae are all limited formulae, the lemma is true for atomic Φ. If $\Phi(x_1,\ldots,x_n)$ is $\psi_1(x_1,\ldots,x_n) \vee \psi_2(x_1,\ldots,x_n)$ and the classes K_{ψ_1} and K_{ψ_2} satisfy the lemma for ψ_1 and ψ_2 respectively, then $K_\Phi = K_{\psi_1} \cup K_{\psi_2}$ is the required class for Φ. If Φ is $\bigvee_y \Psi(y,x_1,\dots,x_n)$ and if K_ψ satisfies the lemma for $\psi(x_0,x_1,\dots x_n)$, then $\{\langle p,z\rangle ; \bigvee_y \langle p,y,z\rangle \in K_\psi\}$ is the required class. The case that Φ is $\bigvee_y^\alpha \Psi(y,x_1,\dots,x_n)$ is similar to the previous one. If Φ is $\neg\,\Psi(x_1,\dots,x_n)$ and if K_ψ satisfies the lemma for Ψ, then

$$\{\langle p,z\rangle ; p \in P \wedge \neg \bigvee_{q \in P}(p \leqslant q \wedge \langle q,z\rangle \in K_\psi)\}$$

is the required class K_Φ. This proves lemma I.

Definition: A set $\mathcal{K}$ of conditions is dense (cofinal) in $\langle P,\subseteq\rangle$, the set of all conditions, iff for every $p \in P$ there is a $q \in \mathcal{K}$ such that $p \leqslant q$.

Definition: A sequence $\mathcal{R}$ of conditions is complete iff $\mathcal{R}$ is well-ordered by $\subseteq$ and of ordertype ω, $\mathcal{R} = \{p^{(0)},p^{(1)},\dots, p^{(k)},\dots\}$, such that $\mathcal{R} \cap \mathcal{K} \neq \emptyset$ for every dense set $\mathcal{K}$ of conditions.

Remark. Both definitions above are given in the meta-language (and not in $\mathfrak{M}$). The notion of a dense subset of a partially ordered set is due to F.Hausdorff who used the name "cofinal". The original definition of "completeness" for sets $\mathcal{R}$ of conditions of P.J.Cohen was a bit more restrictive. The definition given above is due to W.B.Easton (Thesis, Princeton 1964, the main part appeared in the Annals of math.Logic, vol.1(1970)).

Lemma J: If $\mathcal{R}$ is a complete sequence of conditions, then for every $\mathcal{L}$-sentence Φ there exists $p^{(k)} \in \mathcal{R}$ such that $p^{(k)} \Vdash \Phi$.

Proof. Let Φ be given. By lemma I there is a set K in $\mathfrak{M}$ whose elements are just those conditions p for which $p \Vdash \Phi$ holds. By the second extension lemma K is a dense subset of P, thus $K \cap \mathcal{R} \neq \emptyset$, and there are conditions $p \in \mathcal{R}$ such that $p \in K$ and hence $p \Vdash \Phi$, q.e.d.

The following lemma is the only place where we need the countability of $\mathfrak{M}$. Notice that the weaker assumption, namely the $\mathfrak{M}$-set of $\mathfrak{M}$-subsets of P is countable, is already sufficient. This was used e.g. by R.Solovay in the construction of a model $\mathfrak{N}$ which contains a non-constructible Δ_3^1-set of reals.

Lemma K: There are complete sequences of conditions. Moreover, for every condition p there is a complete sequence $\mathcal{R}$ in which p occurs as first element.

Proof. Since $\mathfrak{M}$ is countable (in the meta-theory), there is an enumeration of all sets of $\mathfrak{M}$ and in particular an enumeration of the set of all subsets of P which are in $\mathfrak{M}$. Let $\{s_n; n \in \omega\}$ be an enumeration of the powerset of P in the sense of $\mathfrak{M}$. Take any condition p and define $p^{(0)} = p$. If $p^{(n)}$ is defined, let $p^{(n+1)}$ be any condition in s_n which extends $p^{(n)}$ if such an element exists, otherwise put $p^{(n+1)} = p^{(n)}$. We show that the so-defined sequence $\mathcal{R} = \{p^{(0)}, \ldots, p^{(n)}, \ldots\}$ intersects every dense set of conditions. If $\mathcal{H}$ is a dense set, then it has a number, say n, in the enumeration, thus $\mathcal{H} = s_n$. By definition $p^{(n+1)} \in s_n$ and $p^{(n+1)} \in \mathcal{R}$, q.e.d.

Definition: Let $\mathcal{H}$ be any collection (in the sense of the meta-language) of conditions and let Φ be an $\mathcal{L}$-sentence. We write $\mathcal{H} \Vdash \Phi$ for $(\exists p \in \mathcal{H})(p \Vdash \Phi)$ and similarly $\mathcal{H} \Vdash^* \Phi$ for $(\exists p \in \mathcal{H})(p \Vdash^* \Phi)$. Notice, that if $\mathcal{R}$ is a complete sequence of conditions, then $\mathcal{R} \Vdash \Phi$ and $\mathcal{R} \Vdash^* \Phi$ are equivalent.

Lemma L: Let $\mathcal{R}$ be a complete sequence of conditions and $\Phi(x_1, \ldots, x_n)$ an $\mathcal{L}$-formula. If $\mathcal{R} \Vdash u_1 = v_1, \ldots, \mathcal{R} \Vdash u_n = v_n$ for constant terms $u_1, \ldots, u_n, v_1, \ldots, v_n$, then
$\mathcal{R} \Vdash \Phi(u_1, \ldots, u_n) \Leftrightarrow \mathcal{R} \Vdash \Phi(v_1, \ldots, v_n)$.

This follows by induction on the length of Φ from lemmata F, G and corollary H.

So far we have investigated several useful properties of the forcing relation. In the next section we shall show that every complete sequence of conditions gives raise to a valuation of the predicates $\mathring{a}_j$ so that the resulting sets are generic.

B) COHEN - GENERIC SETS

We shall use the terminology and formalism introduced in section A.

Definition: Let $\mathcal{R}$ be a complete sequence of conditions. Define the function $val_{\mathcal{R}}$ (valuation or interpretation with respect to $\mathcal{R}$) on the set T of all constant terms of the language $\mathcal{L}$ by induction on their degree as follows:

$$val_{\mathcal{R}}(u) = \{val_{\mathcal{R}}(v);\ v \in T \ \&\ \delta(v) < \delta(u) \ \&\ \mathcal{R} \Vdash v \varepsilon u\}$$

Finally define:

$\mathfrak{N}_{\mathcal{R}} = \{val_{\mathcal{R}}(u); u \in T\}$
(we shall usually omit the subscript $\mathcal{R}$ from $val_{\mathcal{R}}$ and $\mathfrak{N}_{\mathcal{R}}$).

Lemma M: Let u and v be constant terms. If $p \Vdash u \, \varepsilon \, v$ then there is a constant term w such that $\delta(w) \leq \delta(u)$, $\delta(w) < \delta(v)$ and $p \Vdash u \simeq w$, $p \Vdash w \, \varepsilon \, v$.
(for a proof see e.g. A.Lévy [51] p.141).

Lemma N: $\mathcal{R} \Vdash u = v \Leftrightarrow \mathcal{R} \Vdash u \simeq v \Leftrightarrow val_{\mathcal{R}}(u) \triangleq val_{\mathcal{R}}(v)$.

Lemma O: $\mathfrak{N}_{\mathcal{R}}$ is a transitive set. For each $x \in \mathfrak{M}$, $val_{\mathcal{R}}(\underline{x}) = x$, hence $\mathfrak{M} \subseteq \mathfrak{N}_{\mathcal{R}}$.

Proof: The transitivity of $\mathfrak{N}_{\mathcal{R}}$ follows directly from the definitions of $val_{\mathcal{R}}$ and $\mathfrak{N}_{\mathcal{R}}$. $val(\underline{x}) = x$ follows easily by induction on $\delta(\underline{x})$ using the definition of $val(\underline{x})$, the forcing-definition and the lemma N. Thus the witnessing constants $\underline{x}$ ensure that $\mathfrak{M}$ is contained in $\mathfrak{N}$ as a transitive submodel.

The semantics of $\mathcal{L}$. For each $x \in \mathfrak{N}$ let $r(x)$ be the least $\delta(w)$ for which $val(w) = x$. Thus $x, y \in \mathfrak{N}$ & $x \in y \Rightarrow r(x) < r(y)$ by lemma N. Now the formulae of $\mathcal{L}$ can be interpreted in $\mathfrak{N}$ in the following way:

(i) A term u is interpreted in $\mathfrak{N}$ by val(u).
(ii) $u \, \varepsilon \, v$ holds in $\mathfrak{N}$ iff $val(u) \in val(v)$.
(iii) The sentential connectives $\neg$, $\vee$ and the existential quantifier $\bigvee$ are interpreted as usual by $\sim$, $\dot{\vee}$ and $\exists$.
(iv) $\bigvee_x^{\alpha} \Phi(x)$ holds in $\mathfrak{N}$ iff there exists $y \in \mathfrak{N}$ with $r(y) < \alpha$ such that $\Phi(y)$.
(v) $y_1, \ldots, y_{n_i}$ satisfy $\pi_i(x, \ldots, x_{n_i})$ in $\mathfrak{N}$ iff there are $j_1, \ldots, j_{n_i} \in A$ such thet $\langle j_1, \ldots, j_{n_i} \rangle \in R_i$ and $y_1 = val(\dot{a}_{j_1})$, $\ldots, y_{n_i} = val(\dot{a}_{j_{n_i}})$.

One of the most important steps in showing that $\mathfrak{N}$ is a model of ZF is by proving that $\mathfrak{N}$ can be described to a large extent within $\mathfrak{M}$. When one is dealing with inner models $\mathfrak{M}_1$ of some structure $\mathfrak{M}_2$ in the verification of the axioms in $\mathfrak{M}_1$ one usually uses the fact that $\mathfrak{M}_1$ can be described entirely within $\mathfrak{M}_2$, thus reducing the validity of some axioms in $\mathfrak{M}_1$ to the validity in $\mathfrak{M}_2$. Cohen-extensions $\mathfrak{N}$ of countable ZF-models $\mathfrak{M}$ have similar features. Though $\mathfrak{N}$ extends $\mathfrak{M}$, $\mathfrak{N}$ can be described to

a good extend in $\mathfrak{M}$, so that again questions about the validity of statements in $\mathfrak{N}$ can be reduced to questions which can be posed (and "answered") in $\mathfrak{M}$.
This is the content of the following lemma:

Lemma P: Let Φ be an $\mathcal{L}$-sentence. Then Φ holds in $\mathfrak{N}_{\mathcal{R}}$ iff $\mathcal{R} \Vdash \Phi$.

For a proof see e.g. A.Lévy [51] p.144 or Easton's thesis [14].

Lemma Q: $p \Vdash^{*} \Phi$ iff $\mathfrak{N}_{\mathcal{R}} \models \Phi$ for all complete sequences $\mathcal{R}$ containing p.

Proof (see Lévy or Easton, loc.cit.).

Lemma R: $val(E^{\alpha}x\Phi(x))$ is the set of all members y of $\mathfrak{N}$ such that $r(y) < \alpha$ and y satisfies $\Phi(x)$ in $\mathfrak{N}$. $val(\dot{a}_j) = a_j$ is the set of all $y \in \mathfrak{N}$ such that $r(y) < \lambda$ and $y \in \dot{a}_j$ holds in $\mathfrak{N}$.

Lemma S: For every $j \in A$, $val(\dot{a}_j) \subseteq S$ and $val(\dot{a}_j) \notin \mathfrak{M}$, thus the sets $val(\dot{a}_j) = a_j$ are "new".

Proof. $a_j \subseteq S$ follows easily from clause (3) of the forcing definition and lemma R. We have to prove that a_j is "new". Suppose a_j is not new. Then $a_j \in \mathfrak{M}$ and since $a_j = x$ for some $x \in \mathfrak{M}$, $val(\underline{x}) = x$ by lemma O. Hence $val(\underline{x}) = x = a_j = val(\dot{a}_j)$ and lemma N yields: $\mathcal{R} \Vdash \dot{a}_j \simeq \underline{x}$. Therefore $p \Vdash \dot{a}_j = \underline{x}$ for some $p \in \mathcal{R}$. Since p is finite there are sets $s \in S$ such that p is not defined for $\langle s,j \rangle$. Now define $q_0 = p \cup \{\langle\langle s,j\rangle,0\rangle\}$ and $q_1 = p \cup \{\langle\langle s,j\rangle,1\rangle\}$. q_0 and q_1 are extensions of p. Proceed by cases. If $s \notin x$ then $s \notin val_{\mathcal{R}}(\underline{x}) = x$ for every complete sequence $\mathcal{R}$. Also, if $q_1 \in \mathcal{R}$, then $\mathcal{R} \Vdash s \in \dot{a}_j$ and hence $s \in val_{\mathcal{R}}(a_j) = a_j$ by lemma R (since $s \in S \rightarrow \rho(s) < \lambda$ and $\delta(\dot{a}_j) = \lambda$). Thus $s \in \dot{a}_j$ holds in $\mathfrak{N}_{\mathcal{R}}$ for $\mathcal{R}$ containing q_1 while $s \in \underline{x}$ fails in $\mathfrak{N}_{\mathcal{R}}$. Hence $\mathcal{R} \Vdash^{*} \dot{a}_j \neq \underline{x}$ for every complete sequence $\mathcal{R}$ containing q_1 (by lemma Q). Lemma A (iii) yields: $q_1 \Vdash \dot{a}_j \neq \underline{x}$.
If $s \in x$ then proceed as above and obtain $q_0 \Vdash \dot{a}_j \neq \underline{x}$.
Thus we have shown that every condition p has an extension q such that $q \Vdash \dot{a}_j \neq \underline{x}$. By lemma A (i),(iii) this implies that every condition p forces $\dot{a}_j \neq \underline{x}$. It follows now from lemmata O and P that $a_j = val(\dot{a}_j) \notin \mathfrak{M}$, q.e.d.

Lemma T: $\mathfrak{M}$ and $\mathfrak{N}$ have the same ordinals.

Proof. Since $\mathfrak{M} \subseteq \mathfrak{N}$ by lemma 0 every ordinal of $\mathfrak{M}$ is an ordinal of $\mathfrak{N}$ (notice that since $\mathfrak{M}$ and $\mathfrak{N}$ are transitive $\in$-models the notion of "being an ordinal" is absolut). Now let α be an ordinal of $\mathfrak{N}$. Since $\alpha = \rho(\alpha) \leq \delta(\alpha)$, $\delta(\alpha) \in \mathfrak{M}$, the transitivity of $\mathfrak{M}$ implies $\alpha \in \mathfrak{M}$. Here $\rho(\alpha) \leq \delta(\alpha)$ holds since $s, y \in \mathfrak{N}$ and $x \in y \Rightarrow r(x) < r(y)$, thus $x \in \mathfrak{N} \Rightarrow \rho(x) \leq r(x)$.
But by definition of $r(x)$ we have $r(x) \leq \delta(x)$, thus $\rho(x) \leq \delta(x)$ for all $x \in \mathfrak{N}$, q.e.d.

Having proved all these various lemmata we are able to prove the main-theorem of forcing manely that the structure $\mathfrak{N}$ is a model of ZF.

HAUPTSATZ of the forcing technique

> Let $\mathfrak{M}$ be a countable transitive $\in$-model of ZF and let $\mathfrak{A} = \langle A, R_i \rangle_{i \in I}$ be a (1st-order) relational system in $\mathfrak{M}$. If the forcing relation $\Vdash$ is defined as in section A, then for every complete sequence $\mathcal{R}$ the structure $\mathfrak{N}_{\mathcal{R}}$ is a countable, transitive $\in$-model of ZF which extends $\mathfrak{M}$.

Proof: That $\mathfrak{N}_{\mathcal{R}}$ is transitive and includes $\mathfrak{M}$ has been shown in lemma 0.
Ad.axiom(0): Since $val(\underline{\emptyset}) = \emptyset \in \mathfrak{N}$, the axiom of null-set holds in $\mathfrak{N}$.
Ad(I): Extensionality follows from the transitivity of $\mathfrak{N}$.
Ad(II): If a and b are sets of $\mathfrak{N}$, then $a = val(t_1)$, $b = val(t_2)$ for terms t_1, t_2 of $\mathcal{L}$. Let $\delta(t_1) = \alpha_1$, $\delta(t_2) = \alpha_2$ and $\alpha = \max\{\alpha_1, \alpha_2\} + 1$, then $t_3 = E^{\alpha}x(x \simeq t_1 \vee x \simeq t_2)$ is a term of degree α and $val(t_3)$ is the unordered pair of a and b.
Ad(III): Similar to (II). Ad(IV): Since $val(\underline{\omega}) = \omega$ by lemma 0, the axiom of infinity holds in $\mathfrak{N}$.
Ad(V): The power-set axiom holds in $\mathfrak{N}$: (see: COHEN [11] p.46-47)
Cohen's proof that the power-set axiom holds in the extension $\mathfrak{N}$ (see [10] part II) follows the proof of Gödel [25] that V = L implies the GCH. We follow, instead, an elegant proof due to R.M.Solovay which avoids Gödel's argument (see Easton's thesis [14]).

We shall show, that for any constant term t there exists an ordinal α such that if $val(s) \subseteq val(t)$ holds in $\mathfrak{N}$ (for some constant term s), then $val(s) = val(s^*)$ for some constant term

s^* of degree less than α. Then the power-set of val(t) in the sense of $\mathfrak{M}$ is given by

$$\mathrm{val}(E^{\alpha}x(x \subseteq t)).$$

Let t be a constant term. For each constant term s define (p is any condition):

$$\phi(s) = \{\langle p,w\rangle ; \delta(w) < \delta(t) \wedge p \Vdash w \in s\}$$

By lemma I each $\phi(s)$ is a set of $\mathfrak{M}$ and the correspondence $s \mapsto \phi(s)$ is a function of $\mathfrak{M}$. Let this function be represented by the classterm G. We claim that $\phi(s_1) = \phi(s_2)$ implies $\mathrm{val}(t) \cap \mathrm{val}(s_1) = \mathrm{val}(t) \cap \mathrm{val}(s_2)$.

<u>Proof</u>. Suppose $\phi(s_1) = \phi(s_2)$ and $\mathrm{val}(w) \in \mathrm{val}(t) \cap \mathrm{val}(s_1)$. Then by the definition of the valuation val $\hat{=}$ $\mathrm{val}_{\mathcal{R}}$: $p^{(n)} \Vdash w \in t$ and $p^{(n)} \Vdash w \in s_1$ for some $p^{(n)}$ in the complete sequence $\mathcal{R}$. Further w may be taken so that $\delta(w) < \delta(t)$. Thus $\langle p^{(n)},w\rangle \in \Phi(s_1)$. Since $\phi(s_1) = \phi(s_2)$ we get $\langle p^{(n)},w\rangle \in \phi(s_2)$ and this entails $p^{(n)} \Vdash w \in s_2$. We conclude that $\mathrm{val}(w) \in \mathrm{val}(t) \cap \mathrm{val}(s_2)$ [more precisely : $p^{(n)} \Vdash w \in s_2$ implies by lemma M the existence of a term w^* such that $\delta(w^*) < \delta(s_2)$, $p^{(n)} \Vdash w \simeq w^*$ and $p^{(n)} \Vdash w^* \in s_2$. By lemma N: $\mathrm{val}(w) = \mathrm{val}(w^*)$, thus $\mathrm{val}(w^*) \in \mathrm{val}(s_2)$ implies $\mathrm{val}(w) \in \mathrm{val}(s_2)$. Further, since $\mathrm{val}(w) \in \mathrm{val}(t)$ we get $\mathrm{val}(w) \in \mathrm{val}(t) \cap \mathrm{val}(s_2)$ as stated above].

Thus we have shown that $\mathrm{val}(t) \cap \mathrm{val}(s_1) \subseteq \mathrm{val}(t) \cap \mathrm{val}(s_2)$. The inverse $\supseteq$ follows in the same way, and we have proved that $\phi(s_1) = \phi(s_2)$ implies that $\mathrm{val}(s_1)$ and $\mathrm{val}(s_2)$ are equal modulo val(t).

Let $T_{\delta(t)}$ be the set (in $\mathfrak{M}$) of terms of degree less than $\delta(t)$. Then $\phi(s) \subseteq P \times T_{\delta(t)}$, where P is the set of all conditions. For every $k \subseteq P \times T_{\delta(t)}$ let (T is the class of all constant terms)

$$\psi(k) = \{s;\ s \in T \wedge \phi(s) = k \wedge \bigwedge_{s_1} [s_1 \in T \wedge \phi(s_1) = k \rightarrow \delta(s) \leq \delta(s_1)]\}.$$

($\psi(k)$ may be empty but in any case $\psi(k)$ is a set of $\mathfrak{M}$). By the axioms of powerset and replacement in $\mathfrak{M}$, $\{\gamma;\ \bigvee_k \bigvee_s (k \subseteq P \times T_{\delta(t)} \wedge s \in \psi(k) \wedge \gamma = \delta(s))\}$ is a set of $\mathfrak{M}$. Let β_0 be the supremum of the ordinals of this set and define $\beta_1 = \beta_0 + 1$ and $\alpha = \beta_1 + 1$.

Now suppose that x and y are sets of $\mathfrak{N}$ such that $y = \mathrm{val}(t)$ and $x \subseteq y$ holds in $\mathfrak{N}$. Then $x = \mathrm{val}(u)$ for some constant term u and $\mathrm{val}(u) \subseteq \mathrm{val}(t)$. Thus $\phi(u) = k \subseteq P \times T_{\delta(t)}$ and $\psi(k)$ is not empty, since k is represented as a $\phi(u)$ for some constant term u and

$\psi(k)$ contains those terms of minimal degree. Hence let s_0 be some element of $\psi(k) = \psi(\phi(u))$. Then

$$\mathrm{val}(u) = \mathrm{val}(t) \cap \mathrm{val}(u) = \mathrm{val}(t) \cap \mathrm{val}(s_0).$$

Define $s_1 = E^{\beta_1}x(x \in t \wedge x \in s_0)$. Since $\delta(t) \leq \beta_0 < \beta_1$ and $\delta(s_0) < \beta_1$, s_1 is a constant term of degree β_1 and $\mathrm{val}(u) = \mathrm{val}(s_1)$ is true in $\mathfrak{N}$ (by lemmata M, N and R). Moreover $\delta(s_1) = \beta_1 < \beta_1 + 1 = \alpha$. Thus we have obtained an ordinal α (of $\mathfrak{M}$) with the required properties, q.e.d.

Notice that the proof given above is similar to the proof that the powersetaxioms holds in L - see page 28. Also the proof that the replacementaxiom holds in $\mathfrak{N}$ will be inspired by the corresponding proof for L. We need two lemmata.

<u>Lemma U</u>: Let $\Phi(x_0, x_1, \ldots, x_n)$ be an unlimited formula of $\mathcal{L}$. Then for every ordinal α of $\mathfrak{M}$ there is an ordinal β of $\mathfrak{M}$ such that for every condition p and constant terms $t_1, \ldots, t_n$ of rank less that α,

$$p \Vdash \bigvee_{x_0} \Phi(x_0, t_1, \ldots, t_n) \Leftrightarrow p \Vdash \bigvee^{\beta}_{x_0} \Phi(x_0, t_1, \ldots, t_n).$$

<u>Proof</u>. Let Φ be given and suppose that Φ has no free variables other than $x_0, \ldots, x_n$. By lemma I (see section A) there exists in $\mathfrak{M}$ a class K whose elements are the n+2-tuples $\langle p, t_0, t_1, \ldots, t_n\rangle$ such that $p \Vdash \Phi(t_0, \ldots, t_n)$. Hence

$$C = \{\langle p, t_0, \ldots, t_n\rangle ;\ p \Vdash \Phi(t_0, \ldots, t_n) \wedge \delta(t_i) < \alpha \text{ for } 1 \leq i \leq n\}$$

is also a class of $\mathfrak{M}$. By the axiom of foundation in $\mathfrak{M}$ the following collection C^* is a <u>set</u> of $\mathfrak{M}$:

$$C^* = \{\langle p, t_0, \ldots, t_n\rangle ; \langle p, t_0, \ldots, t_n\rangle \in C \wedge [p \Vdash \bigvee_{x_0} \Phi(x_0, t_1, \ldots, t_n) \rightarrow \\ \rightarrow \bigwedge_{\langle p, t_0^*, t_1, \ldots, t_n\rangle \in C} (\delta(t_0) \leq \delta(t_0^*))]\}$$

Thus C^* contains only those n+2-tuples $\langle p, t_0, \ldots, t_n\rangle$ from C for which t_0 has minimal degree whenever $\bigvee_{x_0} \Phi(x_0, t_1, \ldots, t_n)$ is forced by p. By the replacement axiom in $\mathfrak{M}$,

$$D = \{\gamma;\ \bigvee_{\langle p, t_0, \ldots, t_n\rangle \in C^*} (\gamma = \delta(t_0))\}$$

is again a set of $\mathfrak{M}$, and using again the replacement axiom in $\mathfrak{M}$, there exists an ordinal β such that $\gamma \in D \rightarrow \gamma < \beta$. Then it is easily seen that the equivalence stated in the lemma holds for this β, q.e.d.

<u>Lemma V</u>: Let $\Phi(x_1, \ldots, s_n)$ be an unlimited formula of $\mathcal{L}$ and α be an ordinal of $\mathfrak{M}$. Then there exists a limited formula $\Phi^{\nabla}(x_1, \ldots, x_n)$ such that

$$\bigwedge^{\alpha}_{x_1}\ldots\bigwedge^{\alpha}_{x_n}[\Phi(x\ ,\ldots,x_n)\leftrightarrow\Phi^{\nabla}(x_1,\ldots,x_n)]$$

holds in $\mathfrak{N}$.

<u>Proof</u> by induction on the length of Φ.

<u>Case 1</u>. If Φ is atomic, then we can let Φ^{∇} be Φ.

<u>Case 2</u>. Φ is $\neg\Psi(x_1,\ldots,x_n)$. By the induction hypothesis there is a limited formula $\Psi^{\nabla}(x_1,\ldots,x_n)$ such that $\bigwedge^{\alpha}_{x_1}\ldots\bigwedge^{\alpha}_{x_n}[\Psi(x_1,\ldots,x_n)\leftrightarrow\Psi^{\nabla}(x_1,\ldots,x_n)]$ is true in $\mathfrak{N}$. Hence, we can let Φ^{∇} be $\neg(\Psi^{\nabla})$.

<u>Case 3</u>. Φ is $\bigvee^{\delta}_{y}\Psi(x_1,\ldots,x_n,y)$. Let $\gamma=\max\{\alpha,\delta\}$. By the induction hypothesis there is a limited formula $\Psi^{\nabla}(x_1,\ldots,x_n,y)$ corresponding to Ψ and γ such that:

$$\bigwedge^{\gamma}_{x_1}\ldots\bigwedge^{\gamma}_{x_n}\bigwedge^{\gamma}_{y}[\Psi(x_1,\ldots,x_n,y)\leftrightarrow\Psi^{\nabla}(x_1,\ldots,x_n,y)]$$

holds in $\mathfrak{N}$. It follows that

$$\bigwedge^{\gamma}_{x_1}\ldots\bigwedge^{\gamma}_{x_n}[\bigvee^{\delta}_{y}\Psi(x_1,\ldots,x_n,y)\leftrightarrow\bigvee^{\delta}_{y}\Psi^{\nabla}(x_1,\ldots,x_n,y)]$$

is also true in $\mathfrak{N}$. This shows that we can define Φ^{∇} to be $\bigvee^{\delta}_{y}\Psi^{\nabla}(x_1,\ldots,x_n,y)$.

<u>Case 4</u>. Φ is $\bigvee_{y}\Psi(x_1,\ldots,x_n,y)$. By lemma U there is for given Φ and α an ordinal β such that for each condition p and constant terms $t_1,\ldots,t_n$ of degree less than α,

$$p\Vdash\bigvee_{y}\Psi(t_1,\ldots,t_n,y)\Leftrightarrow p\Vdash\bigvee^{\beta}_{y}\Psi(t_1,\ldots,t_n,y).$$

This means by lemma A, (vii) of section A. that every condition p weakly forces

$$\bigwedge^{\alpha}_{x_1}\ldots\bigwedge^{\alpha}_{x_n}[\bigvee_{y}\Psi(x_1,\ldots,x_n,y)\leftrightarrow\bigvee^{\beta}_{y}\Psi(x_1,\ldots,x_n,y)]$$

Hence also every p in the complete sequence $\mathfrak{R}$ weakly forces this formula, and therefore $\mathfrak{R}$ also strongly forces the formula. By lemma P of this section, this formula holds in $\mathfrak{N}$. Let $\gamma=\max\{\alpha,\beta\}$ and proceed as in case 3 (using β rather than δ), q.e.d.

Using lemma V we are able to prove, that the replacementaxiom (VI) holds in the structure $\mathfrak{N}$. Notice that our proof ressembles very much the proof that (VI) holds in Gödel's model L.

<u>Continuation of the proof for the Hauptsatz</u>

<u>Ad(VI)</u>: The replacement-schema holds in $\mathfrak{N}$. Let t_1 be a constant term of degree α and let $\Phi(x,y)$ be a formula of $\mathcal{L}$ such that it holds in $\mathfrak{N}$ that for every $x\in t_1$ there is precisely one y such that $\Phi(x,y)$. By lemma U there is an ordinal β such that

$$\mathfrak{N}\models\bigwedge^{\alpha}_{x}[\bigvee_{y}\Phi(x,y)\leftrightarrow\bigvee^{\beta}_{y}\Phi(x,y)].$$

By lemma V, there is a limited formula $\phi^{\nabla}(x,y)$ such that

(+) $\mathfrak{N} \models \bigwedge_x^{\zeta} \bigwedge_y^{\zeta} [\phi(x,y) \leftrightarrow \phi^{\nabla}(x,y)]$,

where $\zeta = \max\{\alpha,\beta\}$. It follows from (+) that:

(0) $\mathfrak{N} \models \bigwedge_x^{\alpha} \bigwedge_y^{\beta} [\Phi(x,y) \leftrightarrow \Phi^{\nabla}(x,y)]$.

Let $s = E^{\lambda}y(\bigvee_x^{\alpha} \bigwedge_z^{\beta} y = z \wedge x \in t \wedge \Phi^{\nabla}(x,y))$, then s is a constant term of $\mathcal{L}$, where $\lambda = \max\{\alpha,\beta\} + 1 = \zeta + 1$. It follows that val(s) is the image of val(t) under the function Φ in $\mathfrak{N}$, q.e.d.

Thus we have proved the Hauptsatz.

Digression. What have we done so far? Our main question was whether the independence results we have obtained by means of the Fraenkel-Mostowski-Specker-method for the system ZF^0 (without the axiom of foundation, but assuming the existence of reflexive sets $x = \{x\}$) are also true for ZF = ZF^0 + foundation. Obviously not all independence results carry over to independence results in ZF, since e.g. (AC) and (PW) - see p.62 - are equivalent in ZF while (AC) is independent from (PW) in ZF^0 alone. The general procedure in the construction of a permutation model $\mathfrak{M}$ of ZF^0 was to define some relations R_i $(i \in I)$ between a set A of atoms (i.e. reflexive sets) and then to construct the permutation model $\mathfrak{M}$ over the structure $\mathfrak{A} = \langle A, R_i \rangle_{i \in I}$.

In order to obtain independence results which apply to full ZF-set theory (including foundation) our general idea was to add to a given countable model $\mathfrak{M}$ of ZF a generic copy of a structure $\mathfrak{A} = \langle A, R_i \rangle_{i \in I}$. Obviously, we cannot construct within our meta-theory (which is ZF + (AC)) a model ab ovo, since this would give otherwise a contradiction to Gödel's theorem. But what we can do is to construct from some given model $\mathfrak{M}$ of ZF another model $\mathfrak{N}$ of ZF in which some interesting statements Φ are true while others Ψ fail, thus proving that $\Phi \rightarrow \Psi$ is not derivable from ZF. Again by Gödel's theorem, we have to use the fact that $\mathfrak{M}$ is a ZF-model when proving that the extension $\mathfrak{N}$ is a ZF-model. This we have done by reducing questions about $\mathfrak{N}$ by means of lemma P to questions which can be posed in $\mathfrak{M}$. This is the most astonishing fact, that the extension can be described in the groundmodel $\mathfrak{M}$ (see lemmata I and P). It was the aim of the forcing definition to determine the interior $\in$-structure of the "new" generic sets $a_j (j \in A)$ in such a way that in $\mathfrak{M}$ we have evough information what properties these sets a_j have. These finite amounts of informations were called "conditions". Notice that since we are dealing with finite conditions, the sets a_j generic over $\mathfrak{M}$ determined by these conditions, are called Cohen-generic over $\mathfrak{M}$. The name "Cohen-generic" was chosen in honor of the man who first invented forcing with finite conditions. Forcing with perfect-closed subsets of the real-line is usually called Sacks-forcing and the generic sets obtained by this way of forcing are called Sacks-generic. Forcing with Borel-sets is called Solovay-forcing and the corresponding generic sets Solovay-generic - see Silver's Seminar notes [80] and the articles of Sacks and Solovay.

We have developed Cohen-forcing in a general setting and have obtained for infinite sets S Cohen-generic subsets $a_j \subseteq S$. In many cases we shall take ω as S. The Cohen-generic subsets of ω will be called simply Cohen-generic reals, since every subset of ω determines a real number.

Instead of proving one independence result after the other we shall first collect some additional informations about the generic extensions $\mathfrak{N}$. We have shown (see lemma T) that $\mathfrak{M}$ and the extension $\mathfrak{N}$ have the same ordinals. We ask: do they have the same cardinals? id est: are the ordinals λ which are initial ordinals in the sense of $\mathfrak{N}$ just the initial ordinals of $\mathfrak{M}$? or better: under which conditions is this true? Another question: Under what conditions on $\mathfrak{A} = \langle A, R_i \rangle_{i \in I}$ is the axiom of choice (AC) true in the extension?
Further, what are the conditions $\mathfrak{A}$ has to fullfill in $\mathfrak{M}$ in order to ensure that the extension $\mathfrak{N}$ satisfies the ordering principle? In the following section we discuss these questions and give some solutions.

C) ORDERINGS AND WELLORDERINGS IN GENERIC EXTENSIONS

We start with the presentation of a theorem which says that, if $\mathfrak{M}$ is a countable standard model of ZF + (AC) and $\mathfrak{A}$ is finite, then the model $\mathfrak{N}$ obtained from $\mathfrak{M}$ by adding a generic copy of $\mathfrak{A}$ to $\mathfrak{M}$, satisfies ZF + Axiom of choice.

A necessary remark. In section B we have shown, that the model $\mathfrak{N}$ extends $\mathfrak{M}$ (see lemma O), but $\mathfrak{M}$ need not to be a $\mathfrak{N}$-definable subclass of $\mathfrak{N}$. This, however, can be attained by adding to the forcing language $\mathcal{L}$ a further unary predicate symbol $\dot{g}$. The intended interpretation of $\dot{g}(v)$ is "v is in the groundmodel $\mathfrak{M}$".

The forcing-definition has to be enriched by the clause:

(•) $\quad p \Vdash \dot{g}(t) \Leftrightarrow (\exists x \in \mathfrak{M})(t \triangleq \underline{x})$.

for conditions p and constant terms t. The interpretation of $\mathcal{L}$ in $\mathfrak{N}$, id est the semantics for $\mathcal{L}$, then has to be enriched by:

"y satisfies $\dot{g}$ in $\mathfrak{N}$ iff there is a constant term t such that $y \triangleq val(t)$ and $\mathfrak{a} \Vdash \dot{g}(t)$".

It follows that $\dot{g}$ defines $\mathfrak{M}$ in $\mathfrak{N}$, more precisely, $\{val(\underline{x});\ x \in \mathfrak{M}\}$ is $\mathfrak{N}$-definable by means of $\dot{g}$. Whenever we shall need the fact that $\mathfrak{M}$ is $\mathfrak{N}$-definable we shall assume that forcing was done in a way including clause (•). This assumption is made e.g. in the following theorem, the proof of which is close to Gödel's proof that (AC) holds in L.

Theorem. If $\mathfrak{M}$ is a countable standard model of ZF + (AC) and a is Cohen-generic over $\mathfrak{M}$, then the extension $\mathfrak{N} \triangleq \mathfrak{M}[a]$ is a countable standard model of ZF + (AC).

Proof. (R.B.Jensen [40] p.69). Set up in $\mathfrak{N}$ a ramified language $\mathcal{L}$ with a name $\underline{a}$ for a, names $\underline{x}$ for x in $\mathfrak{M}$ (this is possible in $\mathfrak{N}$, since a and $\mathfrak{M}$ are $\mathfrak{N}$-definable), limited quantifiers $\bigvee^{\alpha}$, limited comprehension operators E^{α} for all ordinals α of $\mathfrak{N}$, so that $\{\langle\alpha, \bigvee^{\alpha}\rangle; \alpha \in On^{\mathfrak{N}}\}$ and $\{\langle\alpha, E^{\alpha}\rangle; \alpha \in On^{\mathfrak{N}}\}$ are classes of $\mathfrak{N}$. $\mathcal{L}$ has furthermore all the symbols of ZF. Obtain by recursion (as usual) the well-formed formulae, so that $\mathcal{L}$, the collection of all these wff's, is a class of $\mathfrak{N}$. Define an interpretation Ω for the constant terms of $\mathcal{L}$ by setting

$$\Omega(\underline{a}) = a,\ \Omega(\underline{x}) = x \quad (\text{for } x \in \mathfrak{M}),$$

and then extending to all constant terms of $\mathcal{L}$ by recursion on δ, the degree (defined here as on p.79). Since the correspondence $x \mapsto \underline{x}$ was by definition $\mathfrak{N}$-definable (see e.g. the conventions on p.79), it follows from the recursion theorem, that the function $\Omega = \{\langle t, \Omega(t)\rangle; t \in T\}$ is $\mathfrak{N}$-definable (T is the class of all constant terms). Let T_α be the set (of $\mathfrak{N}$) of constant terms t of degree less than α. Define $N_\alpha = \{\Omega(t); t \in T_\alpha\}$. It follows that $\bigcup_\alpha N_\alpha$ is the $\mathfrak{N}$-class of all sets of $\mathfrak{N}$.

After these preparatory remarks, let us prove that in $\mathfrak{N}$ every set x can be well-ordered. Let x be any set of $\mathfrak{N}$; then there exists in $\mathfrak{N}$ an ordinal α such that $x \subseteq N_\alpha$.
We claim that T_α can be wellordered in $\mathfrak{N}$. In fact

$K_\alpha = \{\underline{x}, x \in \mathfrak{M} \wedge \delta(\underline{x}) < \alpha\}$ is a set of $\mathfrak{N}$ and included in T_α. Since $M_\alpha = \{\Omega(t), t \in K_\alpha\}$ is a set of $\mathfrak{M}$, this set M_α can be well-ordered in $\mathfrak{M}$ and induces hence (via Ω^{-1}) a well-ordering of K_α. By definition (see p.79) constant terms t of degree less than α are constructed as finite sequences of symbols taken from $S = K_\alpha \cup \{E^\beta; \beta < \alpha\} \cup \{\bigvee^\beta; \beta \leq \alpha\} \cup$ {the ZF-symbols}.
The set of ZF-symbols is countable, hence wellorderable. Thus the set S (of $\mathfrak{N}$) is wellorderable, and the set T_α can be well-ordered, e.g. lexicographically. Let W_α be a well-ordering of T_α. For $y \in N_\alpha$ let t_y by the first $t \in T_\alpha$ (in the ordering W_α) such that $y = \Omega(t)$. The function $\{\langle y, t_y\rangle; y \in x\}$ is $\mathfrak{N}$-definable and hence so is the well-ordering

$$\{\langle y_1, y_2\rangle; y_1, y_2 \in x \wedge \langle t_{y_1}, t_{y_2}\rangle \in W_\alpha\}$$

of x. This proves the theorem.

Corollary 1. Let $\mathfrak{M}$ be a countable, standard model of the NBG-set theory Σ + (E) and let a be Cohen-generic over $\mathfrak{M}$. Then $\mathfrak{N} \triangleq \mathfrak{M}[a]$ is a countable standard model of Σ + (E).

Here Σ is the set of axioms of groups A,B,C,D in Gödel's orange monograph [25] and (E) is the global version of the axiom of choice.

Corollary 2. Let $\mathfrak{M}$ be a countable standard model of ZF + (AC) and $a_1, \ldots, a_n$ be finitely many sets which are Cohen-generic over $\mathfrak{M}$. Then $\mathfrak{N} \triangleq \mathfrak{M}[a_1, \ldots, a_n]$ is a countable standard model of ZF + (AC).

Corollary 2 was obtained by S.Feferman using ideas of Gödel and Cohen:

[16] S.FEFERMAN: Some applications of the notions of forcing and generic sets. Fund.Math. 56(1965)p.325-345. See also Feferman's article (with the same title) in the "Theory of Models"-Symposium volume, North Holland Publ.Comp. Amsterdam 1965.

Symmetry Properties of Generic Extensions. Let $\mathfrak{M}$ be a countable standard model of ZF and let $\mathfrak{A} = \langle A, R_i\rangle_{i \in I}$ be a relational system in $\mathfrak{M}$. Let $\mathfrak{G}$ be the group in $\mathfrak{M}$ of automorphisms of $\mathfrak{A}$. Let $\mathcal{L}$ be the ramified language having constants $\underline{x}$ for each $x \in \mathfrak{M}$, constants $\dot{a}_j$ for each $j \in A$ and n_i-ary predicate symbols π_i for each $i \in I$

(as defined in section A). Let $\Vdash$ be the forcing relation as defined in section A. For $\sigma \in \mathfrak{G}$ and $\mathcal{L}$-formulae Φ let $\sigma(\Phi)$ be the formula obtained from Φ by replacing every occurence of $\dot{a}_j$ in Φ by $\dot{a}_{\sigma(j)}$. For conditions p (id est: finite partial functions from $S \times A$ into $2 = \{0,1\}$, see p.81) define $\sigma(p)$ by:

$$\langle\langle s,j\rangle,0\rangle \in p \Leftrightarrow \langle\langle s,\sigma(j)\rangle,0\rangle \in \sigma(p)$$
$$\langle\langle s,j\rangle,1\rangle \in p \Leftrightarrow \langle\langle s,\sigma(j)\rangle,1\rangle \in \sigma(p)$$

By definition a formula Φ of $\mathcal{L}$ may contain some particular terms like $\underline{x}$ or $E^{\alpha}x\Psi(x)$ but contains never variables for terms. Thus if $E^{\alpha}x\Psi(x)$ occurs in Φ, then $E^{\alpha}x\sigma(\Psi(x))$ occurs in $\sigma(\Phi)$. According to the forcing-relation defined on p.81-82 the following holds:

<u>Symmetry-Lemma</u> (P.J.Cohen).Let Φ be any $\mathcal{L}$-sentence and let p be any condition. Then for every $\sigma \in \mathfrak{G}$ we have $p \Vdash \Phi \Leftrightarrow$ $\Leftrightarrow \sigma(p) \Vdash \sigma(\Phi)$.

<u>Proof</u> by induction on $\mathrm{Ord}(\Phi)$ for limited formulae Φ and then for unlimited Φ by induction on the length of Φ. Exempla gratia, suppose the lemma is true for limited formulae Φ of order $< \alpha$. If Φ has Order α, proceed by cases. If Φ is $u \varepsilon \underline{x}$, then
$p \Vdash u \varepsilon \underline{x} \Leftrightarrow (\exists y \in x)(p \Vdash u \equiv \underline{y}) \Leftrightarrow (\exists y \in x)(\sigma(p) \Vdash \sigma(u) \equiv \underline{y}))$ since $y \in x \Rightarrow \rho(y) < \rho(y)$, hence $\mathrm{Ord}(u \equiv y) < \mathrm{Ord}(u \varepsilon \underline{x})$.
The latter is equivalent to $(\exists y \in x)(\sigma(p) \Vdash \sigma(u) \equiv \underline{y})$ which in turn is by (1) of the forcing definition equivalent to $\sigma(p) \Vdash \sigma(u) \varepsilon \underline{x}$, id est $\sigma(p) \Vdash \sigma(u \varepsilon \underline{x})$. One proceeds similar in all the other cases.

The symmetry-lemma has the following consequence, if $p \Vdash \Phi$ and p is in the complete sequence $\mathfrak{K}$, which defines $\mathfrak{N}$, then by lemma P, Φ holds in $\mathfrak{N}$. If σ is an automorphism of $\mathfrak{A}$ and $\sigma(p) = p$, then $p \Vdash \sigma(\Phi)$, hence $\sigma(\Phi)$ holds in $\mathfrak{N}$ as well. Since $\mathfrak{G}$ is in $\mathfrak{M}$ we can handle symmetry-properties of $\mathfrak{N}$ in the ground model $\mathfrak{M}$.

We shall use the symmetry-lemma in order to prove that there are models $\mathfrak{N}$ of ZF in which choice fails, thus proving Cohen's theorem, that the axiom of choice (AC) is not deducible from ZF. However, we shall not present Cohen's original proof [9], [11]. In the proof given here a Cohen-extension $\mathfrak{N}$ of a countable standard model $\mathfrak{M}$ is constructed in which there exists an infinite, but Dedekind-finite set A. This construction is due to J.D.Halpern-A.Lévy:

[35] J.D.HALPERN-A.LÉVY: The Boolean Prime Ideal theorem does not imply the axiom of choice; Mimeographed Notes, 93 pages. To appear in the Proceedings of the 1967-Set Theory Symposium at UCLA (AMS-Publications).

(see also Jensen's lecture notes [39] p.164-167).

Definitions. A set x is called finite, iff x is equipotent to some member n of ω. A set x is infinite, iff it is not finite. A set x is called Dedekind-finite, iff there does not exist a function f mapping x one-to-one onto some proper subset of x (This definition of finiteness was used in 1888 by R.Dedekind in his famous monograph "Was sind und was sollen die Zahlen"). In ZF it holds obviously that every finite set is Dedekind-finite. In order to prove the converse one needs the axiom of choice; the following fragment of the axiom of choice turns out to be already sufficient:

(AC^{ω}): The countable axiom of choice: For every set x of non-empty sets such that x is countable, there exists a function f such that for all $y \in x$ it holds that $f(y) \in y$.

Lemma: $ZF + (AC^{\omega}) \vdash$ Every infinite set x has a denumerable subset.

Proof. Let x be infinite. Define $S_n = \{y \subseteq x;\ \bar{\bar{y}} = n\}$ for $n \in \omega$. Then $T = \{S_n; 0 < n < \omega\}$ is denumerable. By (AC^{ω}) there exists a function f defined on T such that $f(S_n) \in S_n$ for all n, $1 \leq n < \omega$. Hence $f(S_n)$ contains n elements. Define $g(n) = f(S_n)$ for $1 \leq n < \omega$. The set $\{g(n); 1 \leq n \in \omega\} = G$ is countable. Thus using (AC^{ω}) one obtains a function h defined on G such that $h(g(n)) \in g(n)$. Define $h^*(n) = h(g(n))$, then $\{h^*(n); 1 \leq n < \omega\}$ is an infinite countable subset of x. This set is infinite since x is infinite and therefore every S_n for $1 \leq n < \omega$ non-empty, q.e.d.

Corollary. $ZF + (AC^{\omega}) \vdash$ A set x is finite iff it is Dedekind-finite.

Proof. Let x be Dedekind-finite and suppose x is not finite. Then by the precceding lemma x has a countable infinite subset $y = \{z_1, z_2, z_3, \ldots\}$. Define a function f on y into y by: $f(z_n) = z_{n+1}$, then $f''y = \{z_2, z_3, z_4, \ldots\}$. Extend f to a function f^* defined on the whole of x by $f^*(u) = u$ iff $u \notin y$ and $f^*(u) = f(u)$ iff $u \in y$. Then f^* is a one-to-one mapping from x onto (the proper subset) $x - \{z_1\}$. Thus x would be Dedekind-infinite, a contradiction, q.e.d.

Now we shall construct a Cohen-extension $\mathfrak{N}$ of some ZF-model $\mathfrak{M}$ in which (AC^{ω}) fails by showing that in $\mathfrak{N}$ there are infinite sets which are Dedekind-finite. The model used is due to Halpern-Lévy **[35]** as indicated above.

Theorem. If ZF is consistent, then
"ZF + there exists an infinite set which is Dedekind-finite" is consistent too. Thus (AC^{ω}) is not provable in ZF.

Proof. Let $\mathfrak{M}$ be a countable standard model of ZF. Consider the following structure $\mathfrak{A} = \langle A,R\rangle$ in $\mathfrak{M}$, where A is ω and R is the unary predicate which holds for every $x \in A$; thus $\mathfrak{A} = \langle\omega,\omega\rangle$. Define a ramified language $\mathcal{L}$ in $\mathfrak{M}$ which has besides the usual ZF-symbols also constants $\underline{x}$ for every set x of $\mathfrak{M}$, constants $\dot{a}_i$ for every $i \in \omega = A$, a constant $\dot{b}$ and limited quantifiers $\bigvee^{\alpha}$ and limited comprehension operators E^{α} for all ordinals α of $\mathfrak{M}$, so that the sequences $\{\langle\alpha,\bigvee^{\alpha}\rangle ; \alpha \in On^{\mathfrak{M}}\}$ and $\{\langle\alpha,E^{\alpha}\rangle ; \alpha \in On^{\mathfrak{M}}\}$ are in $\mathfrak{M}$ (see section A for details). Define a condition p to be a partial finite function from $\omega \times \omega$ into $2 = \{0,1\}$ and define the forcing relation $\Vdash$ as in section A. Thus the key-clauses (3) and (7) read in the present context (t is a constant term):

$$p \Vdash t \in \dot{a}_j \Leftrightarrow (\exists n \in \omega)[p \Vdash t \simeq \underline{n} \;\&\; p(\langle n,j\rangle) = 1]$$

$$p \Vdash t \in \dot{b} \Leftrightarrow (\exists j \in \omega)[p \Vdash t \simeq \dot{a}_j].$$

Obtain a complete sequence $\mathcal{R}$ of conditions and thereby an interpretation $val_{\mathcal{R}}$ of the constant terms of the language $\mathcal{L}$, which defines the model $\mathfrak{N}$. Write $a_j = val(\dot{a}_j)$, $b = val(\dot{b})$; then $a_j \subseteq \omega$ for all $j \in \omega$ and $b = \{a_j ; j \in \omega\}$. By our "Hauptsatz", $\mathfrak{N}$ is a model of ZF. We want fo prove that in $\mathfrak{N}$, b is infinite while Dedekind-finite. This is done in several steps.

1 Step. $a_i \neq a_j$ if and only if $i \neq j$.

Proof. Suppose there are integers i and j such that $i \neq j$ and $a_i = a_j$ holds in $\mathfrak{N}$. Then.(by lemma P) $a_i = a_j$ is forced by some p in the complete sequence $\mathcal{R}$, which defines $\mathfrak{N}$. Hence $p \Vdash a_i = a_j$. Since p is finite there exists a natural number n such that $\langle n,i\rangle \notin p$ and $\langle n,j\rangle \notin p$. Since $i \neq j$, we can extend p to a condition q by defining:

$$q = p \cup \{\langle\langle n,i\rangle,1\rangle,\langle\langle n,j\rangle,0\rangle\}.$$

By the forcing definition $q \Vdash n \in \dot{a}_i$ and $q \Vdash \neg\, n \in \dot{a}_j$ where $p \leqslant q$.

Thus, by the 1st Extension-lemma, we have obtained a contradiction.

<u>2.Step</u>. It holds in $\mathfrak{N}$ that $b = \{a_j;\ j \in \omega\}$ is infinite.

<u>Proof</u>. Otherwise there would be a one-to-one function f in $\mathfrak{N}$ mapping b onto some member n of ω. This is impossible by the result proved in the first step.

<u>3.Step</u>. It holds in $\mathfrak{N}$ that b is Dedekind-finite.

<u>Proof</u>. Let f be any one-to-one function in $\mathfrak{N}$ which maps b onto some subset c of b. We claim that $(f''b =) c = b$. It is sufficient to show, that there exists a number $m \in \omega$ such that $f(a_j) = a_j$ for all $j \geqslant m$.

By definition of $\mathfrak{N}$, f is a limited comprehension term $E^\alpha x\Phi(x) = t_f$. By our assumption it holds in $\mathfrak{N}$ that f is one-to-one; thus by lemma P:

$$p \Vdash Fnc(t_f) \wedge Fnc(t_f^{-1}) \wedge Dom(t_f) = \dot{b} \wedge Range(t_f) \subseteq \dot{b}$$

for some p in the complete sequence $\mathfrak{K}$. Let $occ(\Phi)$ be the $\mathfrak{M}$-set of numbers j such that $\dot{a}_j$ occurs in Φ, where $t_f = E^\alpha x\Phi(x)$. Let k be any (sufficient large) natural number such that $occ(\Phi) \subseteq k$ and $Dom(p) \subseteq \omega \times k$. This means: all $j \in occ(\Phi)$ are smaller than k and if $\langle\langle i,n\rangle ,e\rangle \in p$ for some $i \in \omega$, $e \in 2$, then $n < k$.
We claim that $p \Vdash^{*} \langle \dot{a}_j, \dot{a}_j\rangle \in t_f$ for $j \geqslant m = k + 1$.
Otherwise there would exist an extension q of p (by the definition of forcing) and natural numbers n_1, n_2 such that $n_1 \neq n_2$, $m \leqslant n_1$, $m \leqslant n_2$ and

(0) $\quad q \Vdash \langle \dot{a}_{n_1}, \dot{a}_{n_2}\rangle \in t_f$

Choose $h \in \omega$ such that $\langle\langle n,j\rangle ,e\rangle \in q$ implies $n < h$, $j < h$ and such that $n_1 \neq h$, $n_2 \neq h$. Define a permutation σ on ω by $\sigma(h) = n_2$, $\sigma(n_2) = h$, $\sigma(i) = i$ for $i \in \omega - \{h, n_2\}$. An application of the symmetry-lemma to (0) yields:

$$\sigma(q) \Vdash \langle \dot{a}_{n_1}, \dot{a}_h\rangle \in t_f$$

since $occ(\Phi) \subseteq k < m \leqslant n_1, n_2$, hence $\sigma(\Phi) = \Phi$, thus $\sigma(t_f) = \sigma(E^\alpha x\Phi(x)) = E^\alpha x\sigma(\Phi(x)) = E^\alpha x\Phi(x) = t_f$. By definition of σ, $q \cup \sigma(q)$ is a function and hence a condition extending both q and $\sigma(q)$. Therefore by the first extension lemma and lemma B:

$$q \cup \sigma(q) \Vdash \langle \dot{a}_{n_1}, \dot{a}_{n_2}\rangle \in t_f \wedge \langle \dot{a}_{n_1}, \dot{a}_h\rangle \in t_f \wedge Fnc(t_f) \wedge Fnc(t_f^{-1}).$$

Hence $q \cup \sigma(q) \Vdash \dot{a}_{n_2} = t_f(\dot{a}_{n_1}) = \dot{a}_h$, since t_f is a function. This contradicts the result proved in the first step. Thus in fact $p \Vdash^* t_f(\dot{a}_j) = \dot{a}_j$ for all $j \geq m$ and f must be surjectiv. This finishes the proof the theorem.

D) THE POWER OF THE CONTINUUM IN GENERIC EXTENSIONS

If we assume the axiom of choice (AC), then every set x is equipotent with precisely one aleph $\aleph_\alpha$. If $\mathbb{R}$ is the set of all reals then $\overline{\overline{\mathbb{R}}} = \aleph_\nu$ for a certain ordinal ν. Is it possible to determine this ordinal? It follows from Cantors theorem $\bigwedge_x (\bar{\bar{x}} < \overline{\overline{2^x}})$ that $\nu \geq 1$. G.Cantor has spent many years in order to solve this problem without arriving at the determination of the value for ν.

The natural approach to this problem is to determine the cardinalities of various subsets of $\mathbb{R}$. Cantor showed that every perfect set has cardinality $2^{\aleph_0}$ (a set is perfect iff it is a compact subset of $\mathbb{R}$, non-void and every element of it is an accumulation point of it). Moreover, the Cantor-Bendixion-theorem asserts that every closed subset of $\mathbb{R}$ is either countable or the union of a perfect set and a countable set. Thus no closed subset of $\mathbb{R}$ has a cardinal strictly between $\aleph_0$ and $2^{\aleph_0}$. Some further results of classical descriptive set theory read as follows:

(a) Every uncountable $\underset{\sim}{\Sigma}^1_1$-set of reals contains a perfect subset.

(b) Every $\underset{\sim}{\Pi}^1_1$-set of reals is the disjoint union of $\aleph_1$ many Borel sets.

It follows that every $\underset{\sim}{\Sigma}^1_1$-set has power $\leq \aleph_0$ or $= 2^{\aleph_0}$. Since Borel-sets are $\underset{\sim}{\Delta}^1_1$ (Souslin's theorem), hence $\underset{\sim}{\Sigma}^1_1$, it follows that every $\underset{\sim}{\Pi}^1_1$-set of reals has cardinality $\leq \aleph_1$ or $= 2^{\aleph_0}$ (For the notion $\underset{\sim}{\Sigma}^1_1$, etc,... see chapter II, page 44). For a treatment of these results see: [78] and:

[55] A.A.LJAPUNOW: Arbeiten zur deskriptiven Mengenlehre; V.E.B.-Deutscher Verlag der Wissenschaften, Berlin 1955.

Since it was impossible to exhibit a subset of $\mathbb{R}$ of cardinality strictly between $\aleph_1$ and $2^{\aleph_0}$, Cantor conjectured in 1878 that

(CH) $\quad 2^{\aleph_0} = \aleph_1$,

called the "Continuum-Hypothesis". David Hilbert listed this problem as the first problem in his famous list of unsolved problems at the first international congress of Mathematicians in 1900 in Paris. Despite many attempts this problem remained for a long time unsolved. It was however used freely in proofs since it turned out to be a powerful assertion and also often symplified situations. W.Sierpiński deduced a large number of propositions (there called C_1 - C_{82}) from (CH),

[79] W.SIERPIŃSKI: Hypothèse du continu. Warszawa-Lwów 1934
(2nd edition, New York 1956).

In the lit erature there are many papers in which (CH) or the generalized continuum-hypothesis (GCH) is discussed and proved to be equivalent to other statements. W.Sierpiński contributed many papers concerning the (GCH).
H.Rubin has shown, e.g., that the (GCH) is equivalent in ZF to: "For all transfinite cardinals p and q, if p covers q, then for some r it holds that $p = 2^r$".
(see H.Rubin, Bull. AMS.65(1959)p.282-283). B.Sobociński has published a series of notes on the (GCH) in the Notre Dame Journal of formal Logic (parts I, II, III, vol. 3 and 4 (1962,63). K.Gödel has published in 1947 an article in which he gives a survey on results around the (GCH) and in which he discusses the more philosophic problem of the "truth" of the (GCH):

[26] K.GÖDEL: What is Cantor's Continuum Problem? Amer.Math.
Monthly 54(1947)p.515-525, Corrections vol.55(1948)p.151.

Kurt Gödel showed in 1938 that the (GCH) is consistent with ZF, see chapter II of these lecture notes. Thus the (GCH) cannot be refuted in ZF. We have presented here a proof, that (AC) cannot be proved from the ZF-axioms. Since the (GCH) implies the (AC) - see page 24 - it follows, that also the (GCH) is not a theorem of ZF. Thus (GCH) is neither provable nor refutable in the system ZF. But now the following question arises: if we are willing to add the (AC) to the axioms of Zermelo-Fraenkel set theory ZF, is then the continuum-hypothesis (CH) or even the (GCH) derivable? P.J.Cohen [9]-[12] has shown that the (GCH) is not provable in ZF + (AC). Hence the truth or the falsity of the continuum hypothesis cannot be decided on the basis of the usual axioms of set theory, including the axiom of choice.

Theorem (P.J.Cohen). If ZF is consistent, then ZF + (AC) + $2^{\aleph_0} \geqslant \aleph_2$ is consistent too. Thus the continuum hypothesis (CH) is not a theorem of ZF + (AC).

Proof. Let $\mathfrak{M}$ be a countable standard model of ZF + (AC). We shall construct an extension $\mathfrak{N}$ of $\mathfrak{M}$ by adding generically so many new subsets of ω, such that $2^{\aleph_0} = \aleph_1$ is violated in $\mathfrak{N}$. We shall show below that it is sufficient to add $\aleph_2$-in the sense of $\mathfrak{M}$-new subsets of ω to $\mathfrak{M}$. Since $\mathfrak{M}$ is countable, hence has only countably many (in the sense of the meta-language) subsets of ω (though in $\mathfrak{M}$ these sets have cardinality $\geq \aleph_1$), there is hope, that we will find $(\aleph_2)_{\mathfrak{M}}$ - many subsets of ω not yet in $\mathfrak{M}$, since $(\aleph_2)_{\mathfrak{M}}$ is countable outside of $\mathfrak{M}$.

Define in $\mathfrak{M}$ a ramified language $\mathcal{L}$ which has besides the usual ZF-symbols, the limited existential quantifiers $\bigvee^{\alpha}$, the limited comprehension operators E^{α} (for ordinals α in $\mathfrak{M}$), the constants $\underline{x}$ for $x \in \mathfrak{M}$, a further binary predicate symbol $\dot{a}$. Define $\mathcal{L}$ in such a way, so that the correspondences $x \mapsto \underline{x}$, $\alpha \mapsto E^{\alpha}$ and $\alpha \mapsto \bigvee^{\alpha}$ are all classes of $\mathfrak{M}$ (use e.g. the standard trick presented on p.79). Define a condition p to be a finite partial function from $\omega \times \aleph_2$ into $2 = \{0,1\}$. Define the forcing relation $\Vdash$ in the usual way (see page 81-82) containing the following key-clause:

$$p \Vdash \dot{a}(t_1,t_2) \Leftrightarrow (\exists n \in \omega)(\exists \nu \in \aleph_2)(p \Vdash t_1 = \underline{n} \,\&\, p \Vdash t_2 = \underline{\nu} \,\&\, p(\langle n,\nu\rangle) = 1).$$

This means in terms introduced in section **A**: We take as relational system $\mathfrak{A} = \langle A,R\rangle$ the very special case $\mathbf{A} = 1 = \{0\}$ and $R = \emptyset$, and choose a generic copy of $\mathfrak{A}$ in $\mathbf{S} = \omega \times \aleph_2$. Thus by choosing a complete sequence $\mathcal{R}$ of conditions and defining the valuation-function as in section B, our Hauptsatz tells us, that the model $\mathfrak{N}$ obtained in this way is a model of ZF which contains $\mathfrak{M}$ as a submodel and contains $\mathbf{a} = \mathrm{val}(E^{\alpha}\langle x,y\rangle\, \dot{a}(x,y)) \subseteq \omega \times \aleph_2$, for $\alpha = \omega_2^{\mathfrak{M}}$ (the superscript $\mathfrak{M}$ indicates that the concepts are understood in the sense of $\mathfrak{M}$).
By a theorem proved in section C, $\mathfrak{N}$ is also a model of (AC), since we have added to $\mathfrak{M}$, a model of ZF + (AC), only one new Cohen-generic set. Thus it remains to show, that in $\mathfrak{N}$ the continuum hypothesis is wrong.

Since $\aleph_2^{\mathfrak{M}}$ is the ordinal $\omega_2^{\mathfrak{M}}$ in $\mathfrak{M}$ and ordinals are preserved by the transition from $\mathfrak{M}$ to $\mathfrak{N}$, $\omega_2^{\mathfrak{M}}$ is an ordinal of $\mathfrak{N}$. Thus if we define for $\nu < \omega_2^{\mathfrak{M}} = \gamma$:

$$a_{\nu} = \{n;\ n \in \omega \wedge \langle n,\nu\rangle \in \mathbf{a}\}$$

then $a_{\nu} \subseteq \omega$ and $\nu_1 \neq \nu_2 \rightarrow a_{\nu_1} \neq a_{\nu_2}$ (as in the proof of the preceeding theorem) and we get

$$2^{\aleph_0} \geq \overline{\overline{\gamma}}^{\mathfrak{N}} \quad (\text{in } \mathfrak{N}).$$

γ is in $\mathfrak{M}$ the second infinite cardinal: $\gamma = \omega_2^{\mathfrak{M}}$; We shall show that cardinals are preserved in the extension, i.e. an ordinal which is a cardinal in $\mathfrak{M}$ is a cardinal in $\mathfrak{N}$ and vice versa. Then it will follow that $\gamma = \omega_2^{\mathfrak{M}}$ is also the second infinite cardinal in $\mathfrak{N}$: $\gamma = \omega_2^{\mathfrak{N}}$, and hence $2^{\aleph_0} \geq \aleph_2$ in $\mathfrak{N}$ as desired. To this end we need some lemmata.

Lemma 1. If B is in $\mathfrak{M}$ a set of conditions such that its elements are pairwise incompatible, then B is (in $\mathfrak{M}$) countable.

$$(\forall B \subseteq \underline{\text{Cond}})[B \in \mathfrak{M} \ \& \ (\forall p_1, p_2 \in B)(p_1 \neq p_2 \rightarrow p_1 \cup p_2 \notin \underline{\text{Cond}}) \Rightarrow \overline{\overline{B}}^{\mathfrak{M}} \leq \omega].$$

Proof. Cond is the $\mathfrak{M}$-set of all conditions. Suppose the lemma is false, and let B be a set of $\mathfrak{M}$, such that $p_1, p_2 \in B \rightarrow (p_1 = p_2 \vee p_1 \cup p_2 \notin \underline{\text{Cond}})$ and $\overline{\overline{B}}^{\mathfrak{M}} > \omega$. Define $B_n = \{p \in B;\ \overline{\overline{p}} \leq n\}$. Since $\bigcup_{n=1}^{\omega} B_n = B$ and B is uncountable, there is a number $n \in \omega$ such that B_n is (in $\mathfrak{M}$) uncountable.

There are conditions $q \in \underline{\text{Cond}}$ such that $\{p \in B_n;\ q \subseteq p\}$ is in $\mathfrak{M}$ still uncountable, namely the empty condition $q = \emptyset$ has this property. On the other hand the cardinality of all such conditions q is bounded by n, since $q \subseteq p$. Thus we may define m to be the greatest natural number such that there exists a condition q such that $\overline{\overline{q}} = m$ and $\{p \in B_n;\ p \supseteq q\}$ is in $\mathfrak{M}$ uncountable. Let q_0 be such a condition of cardinality m having this property. Now choose in $\{p \in B_n;\ p \supseteq q_0\}$ any condition p_1. Since in B all conditions are pairwise incompatible, the elements of $\{p \in B;\ p \supseteq q_0\}$ are also pairwise incompatible.

$p_1 - q_0$ is not empty, since otherwise $p_1 = q_0$ and p_1 would be included in all conditions in $\{p \in B;\ p \supseteq q_0\}$, and hence compatible with them. Thus we can find $\langle\langle k,\nu\rangle, e\rangle \in p_1 - q_0$ such that $\langle\langle k,\nu\rangle, 1-e\rangle$ is contained in (in the sense of $\mathfrak{M}$) uncountably many conditions from $B^* = \{p \in B_n;\ p \supseteq q_0\}$. This follows, since p_1 is incompatible with every $p \in B^*$. It follows that $\{p \in B_n;\ p \supseteq q_0 \cup \{\langle\langle k,\nu\rangle, 1-e\rangle\}\}$ is uncountable in the sense of $\mathfrak{M}$ and $q_0 \cup \{\langle\langle k,\nu\rangle, 1-e\rangle\}$ has cardinality m+1, a contradiction to the choice of q_0 with maximal cardinality having this property. Thus lemma 1 is proved.

Lemma 2: If f is a function in $\mathfrak{N}$, such that $\text{Dom}(f) \in \mathfrak{M}$ and $\text{Range}(f) \subseteq x$ for some $x \in \mathfrak{M}$, then there exists a function g in $\mathfrak{M}$ such that Dom(f) = Dom(g), Range(f) $\subseteq \bigcup$ Range(g) $\subseteq$ x, and g(s) is in $\mathfrak{M}$ countable for every $s \in$ Dom(f).

Proof. Since $f \in \mathfrak{N}$, there is by definition of $\mathfrak{N}$ a term t_f of the forcing language $\mathcal{L}$ such that $f \triangleq val(t_f)$. Thus the following holds in $\mathfrak{N}$ (for $x,z \in \mathfrak{M}$):

$$(*)\qquad \bigwedge_u \bigwedge_v \bigwedge_w [\langle u,v\rangle \in t_f \wedge \langle u,w\rangle \in t_f \rightarrow v = w] \wedge Dom(t_f) = \underline{z} \wedge Range(f) \subseteq \underline{x}.$$

Since $\mathfrak{N}$ is a generic extension, there is a condition p_0 in the complete sequence $\mathcal{R}$ (which defines $\mathfrak{N}$) such that p_0 forces (*) -see lemma P in section B. Using weak forcing and lemma A of section A, this entails:

$$(**)\qquad (\forall u,v,w \in \mathfrak{M})(\forall q \geq p_0)[q \Vdash^* \langle \underline{u},\underline{v}\rangle \in t_f \;\&\; q \Vdash^* \langle \underline{u},\underline{w}\rangle \in t_f \Rightarrow v = w].$$

Further, for every $u \in$ Dom(f) there is a condition p' in the complete sequence $\mathcal{R}$ such that $p' \Vdash^* \langle \underline{u},\underline{f(u)}\rangle \in t_f$ (this follows since $\langle \underline{u},\underline{f(u)}\rangle \in t_f$ holds in $\mathfrak{N}$). Since both p_0 and p' are in $\mathcal{R}$ and $\mathcal{R}$ is totally ordered by $\subseteq$ we obtain that $p_0 \cup p'$ is a condition. Hence, defining

$$g(s) = \{y;\ y \in x \;\&\; (\exists p' \geq p_0)(p' \Vdash^* \langle \underline{s},\underline{y}\rangle \in t_f)\}$$

for $s \in z$ = Dom(f), we obtain that $f(s) \in g(s)$. The function $g : z \mapsto x$ is in $\mathfrak{M}$ by lemma I of section A, and Dom(g) = Dom(f) = z and Range(f) $\subseteq \bigcup$ Range(g) $\subseteq$ x.

We claim that g(s) is countable in $\mathfrak{M}$ for $s \in z$. For $s \in z$ choose in $\mathfrak{M}$ for every $y \in g(s)$ a condition $p_y \supseteq p_0$ such that $p_y \Vdash^* \langle \underline{s},\underline{y}\rangle \in t_f$. We claim, that $\{p_y;\ y \in g(s)\}$ satisfies the hypothesis of lemma 1. In fact, if $y_1, y_2 \in g(s)$ and $p_{y_1} \cup p_{y_2}$ is a condition, then $p_{y_1} \cup p_{y_2} \Vdash^* \langle \underline{s},\underline{y_1}\rangle \in t_f \;\&\; p_{y_1} \cup p_{y_2} \Vdash^* \langle \underline{s},\underline{y_2}\rangle \in t_f$. But (**) entails $y_1 = y_2$. Hence $p_{y_1} = p_{y_2}$, since for every $y \in g(s)$ we have chosen one p_y. Now lemma 1 yields that $\{p_y; y \in g(s)\}$ is countable. This in turn implies, that g(s) is in $\mathfrak{M}$ countable: $\overline{\overline{g(s)}}^{\mathfrak{M}} \leq \omega$, quod erat demonstrandum.

Notice, that we could interpolate between Range(f) and x only a "multivalued" function g, since the whole complete sequence $\mathcal{R}$ is not in $\mathfrak{M}$, and could thus not be used in order to find the interpolating function g (if $\mathcal{R}$ would be available in $\mathfrak{M}$, we could show $\mathfrak{M} = \mathfrak{N}$, hence $f \in \mathfrak{N}$, but this is contradictory).

But this defect is not too heavy since $g(s)$ is for $s \in z$ always countable, as we have shown.

Lemma 3. Cardinals are absolut in the extension from $\mathfrak{M}$ to $\mathfrak{N}$.

Proof. Let α be an ordinal of $\mathfrak{N}$ (and hence of $\mathfrak{M}$, by lemma T of section B), and let $\overline{\overline{\alpha}}^{\mathfrak{N}}$ be the cardinal of α in $\mathfrak{N}$ and let $\overline{\overline{\alpha}}^{\mathfrak{M}}$ be the cardinal of α in $\mathfrak{M}$ (i.e. the least ordinals equipotent with α). Since $\mathfrak{M} \subseteq \mathfrak{N}$, every function from ordinals $\beta \leqslant \alpha$ onto α which is in $\mathfrak{M}$ is also in $\mathfrak{N}$.
Hence $\overline{\overline{\alpha}}^{\mathfrak{N}} \leqslant \overline{\overline{\alpha}}^{\mathfrak{M}}$. We shall show that also $\geqslant$ holds.

Let f be a function in $\mathfrak{N}$ from $\delta_0 = \overline{\overline{\alpha}}^{\mathfrak{N}}$ onto $\delta_1 = \overline{\overline{\alpha}}^{\mathfrak{M}}$. If α is finite, then $f \in \mathfrak{M}$ and $\delta_0 = \delta_1$ follows trivially. Hence let us assume that α is infinite. By lemma 2 there exists in $\mathfrak{M}$ a function g such that $\mathrm{Dom}(g) = \mathrm{Dom}(f) = \delta_0$ and $\delta_1 = \mathrm{Range}(f) \subseteq \bigcup \mathrm{Range}(g) \subseteq \delta_1$ (hence $=$), and $s \in \delta_1 \to g(s)$ is countable in $\mathfrak{M}$. Hence:

$$\overline{\overline{\alpha}}^{\mathfrak{M}} = \delta_1 = \overline{\overline{\mathrm{Range}(f)}}^{\mathfrak{M}} = \overline{\overline{\bigcup \mathrm{Range}(g)}}^{\mathfrak{M}} \leqslant \overline{\overline{\mathrm{Dom}(g) \times \omega}}^{\mathfrak{M}} = \overline{\overline{\mathrm{Dom}(g)}}^{\mathfrak{M}}$$

(since $\mathrm{Dom}(g) = \mathrm{Dom}(f)$ is infinite and (AC) holds in $\mathfrak{M}$),

$$\overline{\overline{\mathrm{Dom}(g)}}^{\mathfrak{M}} = \overline{\overline{\mathrm{Dom}(f)}}^{\mathfrak{M}} = \overline{\overline{\delta_0}}^{\mathfrak{M}} = (\overline{\overline{\alpha}}^{\mathfrak{N}})^{\mathfrak{M}} = \overline{\overline{\alpha}}^{\mathfrak{N}} .$$

Hence $\overline{\overline{\alpha}}^{\mathfrak{N}} = \overline{\overline{\alpha}}^{\mathfrak{M}}$ and the lemma is proved.

The lemma implies that in particular -- the notion of being $\aleph_2$ (the second infinite cardinal) is absolute in the extension from $\mathfrak{M}$ to $\mathfrak{N}$. Thus the continuum has in $\mathfrak{N}$ power $\geqslant \aleph_2$. This proves the theorem.

Theorem (P.J.Cohen). If ZF is consistent, then so is ZF + (AC) + (GCH) + $V \neq L$. Thus the axiom of constructibility is not provable in ZF + (GCH) + (AC).

Proof. Let $\mathfrak{M}$ be a countable standard model of ZF + $V = L$ and let $\mathfrak{N}$ be the model obtained by adding to $\mathfrak{M}$ one Cohen-generic real a, $\mathfrak{N} = \mathfrak{M}[a]$. Then $V \neq L$ holds in $\mathfrak{N}$ since the class of constructible sets of $\mathfrak{N}$ depends only on the class of ordinals which are in $\mathfrak{N}$. But $\mathfrak{M}$ and $\mathfrak{N}$ have the same ordinals, hence $\mathfrak{M}$ is in $\mathfrak{N}$ the class of constructible sets. Since $a \notin \mathfrak{M}$, we infer that a is not constructible in $\mathfrak{N}$. On the other hand (AC) holds in $\mathfrak{N}$ by the first theorem of ection C. Further the (GCH) holds in $\mathfrak{N}$ since $a \subseteq \omega$, and $V = L(a)$ holds in $\mathfrak{N}$. To see this, use the corresponding proof of $L \models$ (GCH) in chapter II, q.e.d.

The continuum hypothesis can be violated in generic extensions in various ways. R.M.Solovay has extended the result of Cohen and has shown that $2^{\aleph_0}$ can be anything it ought to be. The only values excluded for $2^{\aleph_0}$ are those excluded by König's theorem, which asserts that $2^{\aleph_0}$ is of cofinality greater than $\aleph_0$.

[81] R.M.SOLOVAY: $2^{\aleph_0}$ can be anything it ought to be; In: The Theory of Models, 1963 Symposium at Berkely; North Holland Publ. Comp. Amsterdam 1965, p.435.

Theorem (Solovay [61]): Let $\aleph_\alpha$ be an infinite cardinal in the countable standard model $\mathfrak{M}$ with $\aleph_0 < cf(\aleph_\alpha)$. Then there is an extension $\mathfrak{N}$ of $\mathfrak{M}$ such that the ordinals (cardinals) of $\mathfrak{N}$ are precisely the ordinals (cardinals, resp.) of $\mathfrak{M}$ and $2^{\aleph_0} = \aleph_\alpha$ in $\mathfrak{N}$.

The (GCH) can also be violated in various other forms. E.g. $2^{\aleph_\alpha} = \aleph_{\alpha+1}$ for ordinals $\alpha < \gamma$ and $2^{\aleph_\gamma} \neq \aleph_{\gamma+1}$ (Solovay [81], Derrick-Drake, H.Schwarz et al.), thus answering a problem of Hajnal (Zeitschr.math. Logik u.Gr.Math.vol.2(1956)p.131-136). For a proof of this result see e.g. Jensen [40] p.68-74, the thesis of Schwarz (cited on p.76) and the article of Derrick-Drake in the same volume as Jensen [40].

Solovay has furthermore shown that there are generic extensions in which $2^{\aleph_i} = \aleph_{n_i}$ is consistent where $i < n_i$ and $n_0 \leqslant n_1 \leqslant \ldots \leqslant n_k$ $(k \in \omega)$ is any sequence of natural numbers. W.B.Easton has extended this result. By means of forcing with a proper class of conditions he constructs a Cohen-generic extension $\mathfrak{N}$ of a countable standard model $\mathfrak{M}$ of NBG-set theory (with global choice) in which:

$$2^{\aleph_\alpha} = \aleph_{G(\alpha)} \quad \text{for every regular cardinal } \aleph_\alpha \text{ holds,}$$

where G is any function in $\mathfrak{M}$ from ordinals to ordinals satisfying the following two requirements: (1) $\alpha \leqslant \beta$ implies $G(\alpha) \leqslant G(\beta)$ and (2) $\aleph_{G(\alpha)}$ is not cofinal with any cardinal less than or equal to $\aleph_\alpha$. This result is contained in Easton's thesis (Princeton 1964), published partly in:

[14] :W.B.EASTON: Powers of regular cardinals. Annals of math. Logic, vol.1 (1970),

J.R.Shoenfield has developed a method for obtaining generic extensions of countable standard models of ZF without using ramified languages. Dana Scott told us, that Shoenfield's approach is equivalent with the Boolean-valued model approach (see the forthcoming

article of Scott-Solovay in the UCLA-set theory Symposium proceedings, vol.2).
Shoenfield presents the result of Easton in his article:

[77] J.R.SHOENFIELD: Unramified Forcing; Proceedings of the 1967-set theory symposium at UCLA, to appear in the AMS-publications.

E) THE INDEPENDENCE OT THE (BPI) FROM THE ORDERING - THEOREM

We consider the following statement:

(BPI) Boolean Prime Ideal Theorem: Every Boolean algebra has a prime ideal.

A Boolean algebra B is a distributive, complemented lattice $\mathbb{B} = \langle B, \sqcup, \sqcap, - \rangle$, where $x \sqcup y$ is the join of x and y, $x \sqcap y$ the meet, and -x the complement of x. B can be partially ordered by defining $x \leq y \leftrightarrow x \sqcup y = y$. Then $x \sqcup z$ is the least upper bound for x and z in B, and $x \sqcap y$ is the greatest lower bound for x and y. The maximal element in B is denoted by 1_B and the minimal element by 0_B. An ideal I in $\mathbb{B}$ is a subset of B satisfying the following three conditions:

(i) $0_B \in I$,
(ii) $x \in I \wedge y \leq x \rightarrow y \in I$,
(iii) $x \in I \wedge y \in I \rightarrow x \sqcup y \in I$.

A prime ideal is an ideal with the additional property:

(iv) $x \in I \leftrightarrow (-x) \notin I$.

In a Boolean algebra the prime ideals are just the maximal proper ideals.

The Boolean Prime ideal theorem (BPI) has a considerable number of equivalent forms in several branches of mathematics and in logic, although by far not as many as the axiom of choice (AC). The (BPI) is thus an interesting and natural principle of set theory.

Lemma. The following statements are all equivalent (in ZF) with the Boolean-Prime-Ideal theorem (BPI):

(a) The Stone representation theorem: Every Boolean algebra $\mathbb{B}$ is isomorphic to a field of sets.
(b) The Tychonoff-theorem for T_2-spaces: The product of compact Hausdorff-spaces is compact in the product topology.
(c) In every commutative ring with unit, every proper ideal is included in some prime ideal.

(d) The Stone-Čech compactification theorem.

(e) Alaoglu's theorem: The unit sphere of the adjoint of a Banach-space is a compact Hausdorff space.

(f) In every Boolean algebra, there exists a 2-valued measure.

(g) The principle of consistent choices.

(h) The completeness theorem for 1st-order languages: Let Σ be a set of 1st-order sentences with arbitrary many non-logical constants. If Σ is consistent, then it has a model.

(i) The compactness theorem for 1st-order languages: Let Σ be as in (h). If every finite subset of Σ has a model, then Σ has a model.

For a proof see the following papers: Łos-Ryll Nardzewski: Fund. Math. 38 (1951) and Fund. Math. vol. 41 (1954); D.Scott: Bull.AMS 60 (1954) p. 390, L.Henkin: Bull.AMS. 60 (1954) p. 390; H.Rubin - D.Scott: Bull. AMS. 60 (1954) p.389; R.Sikorski: Boolean Algebras (Springer-Verlag Berlin 1964), Appendix.

We are interested here in one of the consequences of the (BPI), namely the ordering principle:

(OP) Every set x can be totally ordered.

We shall use the notions "totally ordered" and "linearly ordered" synonymously (i.e. equivalently). A somehow stronger principle is the following:

(OE) Order-Extension-Principle: If x is a set and r a partial ordering on x, then there exists a linear ordering t on x such that $r \subseteq t$.

The (OE) has been discovered by Banach, Kuratowski and Tarski (see: W.Sierpiński: Zarys terji mnogosci, Warszawa 1928, p. 158). The first proof which appeared in print is due to E.Marczewski (Szpilrajn) (Fund. Math. 16 (1930) p. 386-389). Marczewski used the lemma of Zorn-Kuratowski in order to deduce (OE). Łos, Ryll-Nardzewski and L.Henkin observed, that (OE) is already a consequence of the (BPI) (proof either via the compactness theorem, or directly using the ultrafilter theorem). Thus we have:

$$ZF \vdash (AC) \rightarrow (BPI) \rightarrow (OE) \rightarrow (OP).$$

We are interested in the problem, whether the converses of these implications also hold. It is not known, whether (OE) → (BPI) is provable in ZF or not. In this section we shall present a result of Adrian R.Mathias, which says, that (OP) → (OE) is not a theorem of ZF. In the next section we shall present the proof of J.D.Halpern-A.Lévy, that (BPI) does not imply the axiom of choice (AC).

Preparatory remarks. Mostowski has constructed in his paper [64] a model $\mathfrak{M}$ containing urelements (atoms) in which the ordering principle (OP) holds while (AC) fails in it. Mostowski takes a countable set of atoms linearly ordered of type η_0 (i.e. the order type of the rationals). A set is called symmetric iff it is mapped onto itself by some finite-support subgroup of G, the group of all order-preserving mappings from η_0 onto η_0. A set x is in the model $\mathfrak{M}$ iff x is hereditarily symmetric. The proof that (OP) holds in $\mathfrak{M}$ is based on the fact, that every set x of $\mathfrak{M}$ has a unique minimal support, supp(x), where supp(x) is a finite subset of η_0 (the set of atoms). The correspondence $x \mapsto \text{supp}(x)$ is in $\mathfrak{M}$ and hence the lexicographic ordering of supp(y) for $y \in x$ together with wellorderings of the sets $K_e = \{y \in x;\ \text{supp}(y) = e\}$ can be used to obtain a (symmetric) totalordering of x.

This idea can be carried over to Cohen-generic extensions. The rôle which was played by the urelements in Mostowski's model $\mathfrak{M}$, will be played by generic reals in the Cohen-extension. However, instead of adding a generic copy of η_0 (the rationals) to some countable standard model $\mathfrak{M}$ of ZF + V = L (in this case we would not know, how to destroy (OE) in the extension $\mathfrak{N}$) we shall add a generic copy of a certain partially ordered set $\langle I,\leqslant\rangle$ to $\mathfrak{M}$. If $\langle I,\leqslant\rangle$ has sufficiently enough automorphisms, then the generic copy of $\langle I,\leqslant\rangle$ will not have in the extension $\mathfrak{N}$ a total ordering, which extends $\leqslant$ (the symmetry-lemma will be used here).

What are the properties, $\langle I,\leqslant\rangle$ has to fulfill, so that in the extension $\mathfrak{N}$ the ordering principle (OP) remains true. The typical property of η_0, which was used by Mostowski in [64] in order to prove the existence of a unique, minimal, finite support of every set x of his model $\mathfrak{M}$ (the supports are sets of urelements!) was the homogeneous ordering of η_0. We shall show, that, if we require that $\langle I,\leqslant\rangle$ is a countable, homogeneous, $\aleph_0$-universal partially ordered set in $\mathfrak{M}$, and $\mathfrak{M} \models$ ZF + V = L, then in $\mathfrak{N}$ (the extension of $\mathfrak{M}$) every set $x = \text{val}(E^\alpha x\Phi(x))$ has a unique, minimal, finite support supp(x), such that the correspondence $x \mapsto \text{supp}(x)$ is $\mathfrak{N}$-definable. The verification of (OP) in $\mathfrak{N}$ is then standard. We need here, that $\mathfrak{M}$ is a model of V = L for two reasons, first in order to establish (using a theorem of B.Jónsson) that there are in $\mathfrak{M}$ homogeneous, universal partially ordered sets, and second in order to ensure that $K_e = \{x;\ \text{supp}(x) \subseteq e\}$, e a finite subset of I, has a definable wellordering in $\mathfrak{N}$. Finally let us notice that, in contrast to Mostowski's permutation model $\mathfrak{M}$, the fact, that we have chosen a partially ordered set $\langle I, \leqslant \rangle$ and not a linearly ordered set (like Mostowski's η_0) will not cause any troubles when we want to linear order lexicographically the supports, since the generic sets a_i (for $i \in I$) are subsets of ω and $\{a_i; i \in I\}$ has thus in $\mathfrak{N}$ a definable ordering (namely the ordering of the real-line).

Having clarified the basic ideas behind the construction of a model $\mathfrak{N}$ of ZF + (OP) + $\neg$ (OE) + $\neg$ (BPI), we start to present the details of the proof. First we define the notion of a homogeneous, universal relational system. The notion is a generalization of Hausdorff's notion of an η_α-set. For more information we refer our reader to the following publications:

[41] B. JONSSON: Homogeneous universal relational systems; Math. Scand. vol.8 (1960) p. 137-142.

[3] J.L.BELL - A.B.SLOMSON: Models and Ultraproducts; North-Holland publ. Comp. Amsterdam 1969. (chapter 10).

Definition. The type τ of a relational system $\mathfrak{A} = \langle A, R_i \rangle_{i \in \zeta}$ where ζ is a finite ordinal) is a sequence $\langle n_0, n_1, \dots n_{\zeta-1} \rangle$ of natural numbers such that for $0 \leqslant i \leqslant \zeta-1$ the relation R_i is n_i-ary.

Definition. The relational system $\mathfrak{B} = \langle B, S_i \rangle_{i \in \zeta}$ is a subsystem of $\mathfrak{A} = \langle A, R_i \rangle_{i \in \zeta}$ iff $B \subseteq A$ and $S_i = R_i \cap B^{n_i}$ (restriction of R_i to B).

Definition. Let $\mathcal{K}$ be a class of relational systems all of the same type τ. A system $\mathfrak{A}$ is $\mathcal{K}$-homogeneous, iff the following holds:

(1) $\mathfrak{A} \in \mathcal{K}$

(2) If $\mathfrak{B} = \langle B, S_i \rangle_{i \in \zeta}$, $\mathfrak{B} \in \mathcal{K}$, and if $\mathfrak{B}$ is a subsystem of $\mathfrak{A} = \langle A, R_i \rangle_{i \in \zeta}$ such that $\bar{\bar{B}} < \bar{\bar{A}}$, and if ϕ is an isomorphism of $\mathfrak{B}$ into $\mathfrak{A}$, then ϕ can be extended to an automorphism of $\mathfrak{A}$.

Definition. Let $\mathcal{K}$ be a class of relational systems all of the same type and let α be an ordinal. A system $\mathfrak{A} \in \mathcal{K}$ is called $(\aleph_\alpha, \mathcal{K})$-universal iff every system $\mathfrak{B} = \langle B, S_i \rangle_{i \in \zeta}$, such that $\bar{\bar{B}} \leqslant \aleph_\alpha$, is isomorphic to a subsystem of $\mathfrak{A}$, and A has itself cardinality $\aleph_\alpha$.

Bjarni Jónsson has proved in [41] under the assupmtion of the (GCH), that under certain conditions on $\mathcal{K}$ and $\aleph_\alpha$ there are $(\aleph_\alpha,\mathcal{K})$-universal, homogeneous relational systems. It follows in particular that there are countable $\aleph_0$-universal, homogeneous partially ordered sets (here $\mathcal{K}$ is the class of all partially ordered sets).

Theorem (A.R.D.Mathias): Let $\mathfrak{M}$ be a countable, standard model of ZF + V = L; then $\mathfrak{M}$ can be extended to a countable standard model $\mathfrak{N}$ of ZF + (OP) + $\neg$ (OE). Thus the orderextension principle (OE) and a fortiori the Boolean prime ideal theorem (BPI) is independent from the ordering principle (OP) in the system ZF.

This result is contained in

[58] A.R.D.MATHIAS: The Order Extension Principle; Proceedings of the 1967-set theory symposium at UCLA. To appear.

Proof. By the theorem of B.Jónsson [41] there exists in $\mathfrak{M}$ a countable, $\aleph_0$-universal, homogeneous parially ordered set $\langle I,\leq\rangle$. We shall extend $\mathfrak{M}$ by adding to $\mathfrak{M}$ a generic copy of $\langle I,\leq\rangle$. We emphasize that if we write $i < j$ then $i \leq j$ and $i \neq j$. Thus $<$ is irreflexive, while $\leq$ is reflexive: $i \leq j \leftrightarrow (i < j \vee i = j)$.

We construct in $\mathfrak{M}$ a ramified language $\mathcal{L}$ with the usual limited quantifiers $\bigvee^\alpha$ and limited comprehension operators E^α (for ordinals α in $\mathfrak{M}$), the ZF-symbols, constants $\underline{x}$ for each set x of $\mathfrak{M}$, individual constants $\dot{a}_i$ for each $i \in I$ and two further constants $\dot{A}$ and $\dot{<}$. The wellformed formulae and limited comprehension terms are defined as usual, with the restriction that if $\dot{A}$ or $\dot{<}$ occurs in the $\mathcal{L}$-formula Φ then $E^\alpha x\Phi(x)$ is a limited comprehension term only if $\alpha \geq \omega + 1$.

A condition p is a finite partial function from $\omega \times I$ into $2 = \{0,1\}$. Define the (strong) forcing relation $\Vdash$ as in section A of this chapter. In our present case clause (3) reads:

$$p \Vdash t \in \dot{a}_i \leftrightarrow (\exists n \in \omega)(p \Vdash \underline{n} = t \;\&\; p(\langle n,i\rangle) = 1)$$

where t is any constant term of $\mathcal{L}$. Clause (7) reads:

$$p \Vdash t \in \dot{A} \leftrightarrow (\exists i \in I)(p \Vdash t = \dot{a}_i)$$

$$p \Vdash t_1 \dot{<} t_2 \leftrightarrow (\exists i_1 \in I)(\exists i_2 \in I)(i_1 < i_2 \;\&\; p \Vdash t_1 = \dot{a}_{i_1} \;\&\; p \Vdash t_2 = \dot{a}_{i_2})$$

Let again $\Vdash^*$ denote the weak forcing relation. It follows

(1) $i_1 < i_2 \Rightarrow \emptyset \Vdash^* \dot{a}_{i_1} \dot{<} \dot{a}_{i_2}$,

(2) $(\forall i \in I)(\emptyset \Vdash^* \dot{a}_i \varepsilon \dot{A})$.

Obtain a complete sequence $\mathfrak{R}$ of conditions and thereby a valuation $val_{\mathfrak{R}}(t)$ of the constant terms t of $\mathcal{L}$, which defines the model $\mathfrak{N}$. Write $a_i = val(\dot{a}_i)$, $A = val(\dot{A})$, $\nabla = val(\dot{<})$, then $a_i \subseteq \omega$,

$A = \{a_i; i \in I\}$ and ∇ is an irreflexive parial ordering (in $\mathfrak{N}$) of A. By our Hauptsatz, $\mathfrak{N}$ is a model of ZF. All what remains to show is, that the ordering principle (OP) holds in $\mathfrak{N}$ while (OE) fails in $\mathfrak{N}$. To this end we need the following restriction lemma and the symmetry - lemma which we have proved in its full generality in section C (see page 99), so that it is available in the present situation.

<u>Restriction lemma:</u> Suppose that $p \Vdash \Phi$ and let $occ(\Phi)$ be the finite set of elements of I such that $i \in occ(\Phi)$ iff $\dot{a}_i$ occurs in Φ. Further let $p/occ(\Phi) = \{\langle\langle n,i\rangle,e\rangle \in p; i \in occ(\Phi)\}$, then $p/occ(\Phi) \Vdash^* \Phi$.

<u>Proof</u>. Define $c_0 = occ(\Phi)$, $d = \{i \in I; \bigvee_{n\in\omega} \bigvee_{e\in 2}(\langle\langle n,i\rangle,e\rangle \in p \wedge i \notin c_0)\}$, and $q = p/occ(\Phi)$. Suppose $\sim q \Vdash^* \Phi$. Then by lemma A (i) (see section A) there exists an extension q' of q such that $q' \Vdash \neg \Phi$. q' mentions names of reals in c_0, and also others, say those in the finite set c_1 :

$$c_1 = \{i \in I; \bigvee_{n\in\omega} \bigvee_{e\in 2}(\langle\langle n,i\rangle,e\rangle \in q' \wedge i \notin c_0)\}.$$

By defintion $c_0 \cap d = c_0 \cap c_1 = \emptyset$. By the univerality of $\langle I,\leqslant\rangle$ we can find a subset c_2 of I, such that $\langle c_0 \cup d,\leqslant\rangle$ and $\langle c_0 \cup c_2,\leqslant\rangle$ are isomorphic, $c_2 \cap (c_0 \cup d) = \emptyset$. and there exists an isomorphism τ which is identical on c_0. By the homogeneity of $\langle I,\leqslant\rangle$, τ can be extended to an automorphism σ of $\langle I,\leqslant\rangle$. Hence: $\sigma(i) = i$ for $i \in c_0$ and $\sigma(j) \in c_2$ for $j \in c_1$. By the symmetry lemma, $q' \Vdash \neg\Phi$ implies $\sigma(q') \Vdash \sigma(\neg\Phi)$. But σ is the identity on $c_0 = occ(\Phi)$, hence $\sigma(\Phi) \triangleq \Phi$, and we obtain $\sigma(q') \Vdash \neg \Phi$. By construction of σ, the domain of the functions p and $\sigma(q')$ coincides only on a (finite subset of $\omega \times c_0$, where both have the same values, since $q \subseteq p$ and $q \subseteq q'$, $q \subseteq \sigma(q')$. Hence $p \cup \sigma(q')$ is a function and therefore a condition. By the first extension lemma (see section A): $p \cup \sigma(q') \Vdash \neg \Phi$. On the other hand $p \Vdash \Phi$ entails also $p \cup \sigma(q') \Vdash \Phi$, a contradiction to the consistency - lemma (see section A). This proves $q \Vdash^* \Phi$, q.e.d.

Lemma: The Orderextension Principle (OE) does not hold in $\mathfrak{N}$. In particular $\nabla = val(\dot{<})$ cannot be extended in $\mathfrak{N}$ to a total ordering of $A = val(\dot{A})$.

Proof. Suppose ∇ can be extended in $\mathfrak{N}$ to a linear ordering of A. Let R be such an orderextension of ∇. By the definition of $\mathfrak{N}$ there exists a limited conprehension term $t \triangleq E^{\alpha}x\Phi(x)$ of $\mathcal{L}$ such that $R = val(t)$. Let $\Psi(t)$ be the $\mathcal{L}$-sentence "t is a totalordering of $\dot{A}$ extending $\dot{<}$" Let $c = occ(t)$ be the set of indices $i \in I$ such that $\dot{a}_i$ occurs in t. Hence $c = occ(\Psi(t))$. Let $S(c) = \{i \in I;\ \bigvee_{j \in c} (i \leqslant j \vee j \leqslant i)\}$ be the "shadow of c in $\langle I, \leqslant \rangle$".
By the universality of $\langle I, \leqslant \rangle$ we may embed the following partially ordered set into $I - S(c)$:

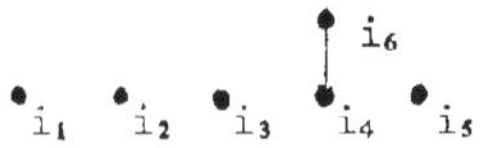

[all elements are pairwise incomparable, only i_4 is smaller than i_6].

Since $R = val(t)$ is a linear ordering of $\dot{A}$ in, it holds in $\mathfrak{N}$ that $\{a_{i_1}, a_{i_2}, a_{i_3}\}$ are ordered by R. Assume e.g. that $\langle a_{i_1}, a_{i_2} \rangle \in R$, $\langle a_{i_2}, a_{i_3} \rangle \in R$. Hence we have obtained six generic terms $\dot{a}_{i_1}, \ldots, \dot{a}_{i_6}$ such that: (i) $\emptyset \Vdash^* \dot{a}_{i_4} \dot{<} \dot{a}_{i_6}$

(ii) $\emptyset \Vdash^* \neg(\dot{a}_{i_m} \dot{<} \dot{a}_{i_k} \vee \dot{a}_{i_k} \dot{<} \dot{a}_{i_m})$
for $m,k = 1,2,\ldots 6$ with $\langle m,k \rangle \neq \langle 4,6 \rangle$,

(iii) $\emptyset \Vdash^* \neg(\dot{a}_{i_m} \dot{<} \dot{a}_j \vee \dot{a}_j \dot{<} \dot{a}_{i_m})$
for $m = 1,2,\ldots,6$ and $j \in c = occ(t)$

(iv) $\mathfrak{N} \models \langle a_{i_1}, a_{i_2} \rangle \in R \wedge \langle a_{i_2}, a_{i_3} \rangle \in R$.

Since everything which holds in $\mathfrak{N}$ must be forced (strongly or weakly) by some condition in the corresponding complete sequence $\mathfrak{R}$, we obtain, that there is a condition p in $\mathfrak{R}$ such that

$$p \Vdash^* \Psi(t) \wedge \langle \dot{a}_{i_1}, \dot{a}_{i_2} \rangle \,\varepsilon\, t \wedge \langle \dot{a}_{i_2}, \dot{a}_{i_3} \rangle \,\varepsilon\, t$$

By the restriction lemma, we may assume that p contains finitely many ordered pairs $\langle\langle n,i \rangle, e \rangle$ (with $n \in \omega$, $e \in 2$) only with $i \in occ(t) \cup \{i_1, i_2, i_3\}$. Define

$$p_1(occ(t),i_1,i_2) = P/_{occ(t) \cup \{i_1,i_2\}}$$

$$p_2(occ(t),i_2,i_3) = P/_{occ(t) \cup \{i_2,i_3\}}.$$

Then by the restriction lemma:

(+) $p_1(occ(t),i_1,i_2) \Vdash^* \Psi(t) \wedge \langle \dot{a}_{i_1}, \dot{a}_{i_2} \rangle \; \varepsilon \; t$

(++) $p_2(occ(t),i_2,i_3) \Vdash^* \Psi(t) \wedge \langle \dot{a}_{i_2}, \dot{a}_{i_3} \rangle \; \varepsilon \; t.$

Now define mappings τ_1 and τ_2 on $occ(t) \cup \{i_1,i_2,i_6,i_5\}$ and on $occ(t) \cup \{i_2,i_3,i_4,i_5\}$, respectively, by:

$\tau_1 : i_1 \mapsto i_6,\; i_2 \mapsto i_5,\; \tau_1$ identical on $occ(t)$,

$\tau_2 : i_2 \mapsto i_5,\; i_3 \mapsto i_4,\; \tau_2$ identical on $occ(t)$;

Hence τ_1 maps $\langle occ(t) \cup \{i_1,i_2\},\leq \rangle$ isomorphically on $\langle occ(t) \cup \{i_5, i_6\},\leq \rangle$ and similar τ_2 maps $\langle occ(t) \cup \{i_2,i_3\},\leq \rangle$ isomorphically on $\langle occ(t) \cup \{i_4,i_5\},\leq \rangle$. By the homogeneity of $\langle I,\leq \rangle$, τ_1 and τ_2 can by extended to automorphisms σ_1 and σ_2, respectively, of $\langle I,\leq \rangle$. Using σ_1 we obtain from (+) by the symmetry-lemma:

(0) $\quad q_1 = p_1(occ(t),i_6,i_5) \Vdash^* \Psi(t) \wedge \langle \dot{a}_{i_6}, \dot{a}_{i_5} \rangle \; \varepsilon \; t$

and using σ_2 from (++):

(00) $\quad q_2 = p_2(occ(t),i_5,i_4) \Vdash^* \Psi(t) \wedge \langle \dot{a}_{i_5}, \dot{a}_{i_4} \rangle \; \varepsilon \; t$

(notice, that the symmetry-lemma also holds with respect to weak forcing $\Vdash^*$). Since $q_1 \cup q_2$ is a condition and $\emptyset \Vdash^* \dot{a}_{i_4} \mathrel{\dot{<}} \dot{a}_{i_6}$ we obtain:

$$q_1 \cup q_2 \Vdash^* \Psi(t) \wedge \dot{a}_{i_4} \mathrel{\dot{<}} \dot{a}_{i_6} \wedge \langle \dot{a}_{i_6}, \dot{a}_{i_5} \rangle \; \varepsilon \; t \wedge \langle \dot{a}_{i_5}, \dot{a}_{i_4} \rangle \; \varepsilon \; t$$

But t extends $\dot{<}$, hence $\langle \dot{a}_{i_4}, \dot{a}_{i_6} \rangle \; \varepsilon \; t$. But $\Psi(t)$ says, that t is a linear ordering on $\dot{A}$, a contradiction! since what we have shown is the following: p is in $\mathcal{R}$, the complete sequence, which defines $\mathfrak{N}$. Define $\mathcal{R}^*$ to be the sequence starting with p/occ(t), having p as its second element, and containing then all conditions q of $\mathcal{R}$ which extend p. Then $\mathcal{R}^*$ defines obviously the same model $\mathfrak{N}$. Since $P/_{occ(t)} \Vdash^* \Psi(t)$ we infer that every complete sequence $\mathcal{R}$ starting with $P/_{occ(t)}$ must force $\Psi(t)$, and $\Psi(t)$ has to hold, hence also in the model $\mathfrak{N}_0$ defined by some complete sequence $\mathcal{R}_0$ which starts with $P/_{occ(t)}$ and has $q_1 \cup q_2$ as second element. But we have just shown, that in this model $val_{\mathcal{R}_0}(t)$ cannot define a total ordering on $val_{\mathcal{R}_0}(\dot{A})$ which extends $val_{\mathcal{R}_0}(\dot{<})$. In this way we have thus obtained a contradiction. This proves the lemma.

Next we want to prove, that in $\mathfrak{N}$ every set can be totally ordered. The idea behind the proof is the following. It can happen that for different $\mathcal{L}$-formula $\Phi_1(x)$ and $\Phi_2(x)$ we have that

$val(E^{\alpha}x\Phi_1(x)) \hat{=} val(E^{\alpha}x\Phi_2(x))$ where $occ(\Phi_1) \neq occ(\Phi_2)$. We want to show that for every x in $\mathfrak{N}$ there is a $\mathcal{L}$-formula $\Psi(x)$ such that $u = val(E^{\alpha}x\Psi(x))$ and

$$occ(\Psi) \subseteq \bigcap \{occ(\Phi);\ u = val(E^{\alpha}x\Phi(x))\}.$$

Then it follows, that every set u of $\mathfrak{N}$ is the valuation of a limited comprehension term $E^{\alpha}x\Psi(x)$ with minimal set $occ(\Psi)$, called the support of u. We then have to show, that the correspondence $u \mapsto$ support of u is definable in $\mathfrak{N}$. Then the rest of the proof that (OP) holds in $\mathfrak{N}$ is standard.

Notation: If $t = E^{\alpha}x\Phi(x)$ is a limited comprehension term and $occ(\Phi) = c$, then we shall write $t \hat{=} t(c)$ in order to indicate that t mentions reals $\dot{a}_i$ if and only if $i \in c$.

Lemma. Let $t(c,d_1)$ and $t'(c,d_2)$ be limited comprehension terms mentioning only reals $\dot{a}_i$ for $i \in c \cup d_1$, $i \in c \cup d_2$ respectively where c,d_1,d_2 are finite disjoint subsets of I. Suppose that

$$\mathfrak{N} \models t(c,d_1) = t'(c,d_2),$$

then there is a limited comprehension term $t'' \hat{=} t''(c)$ mentioning reals $\dot{a}_i$ for $i \in c$ such that

$$\mathfrak{N} \models t(c,d_1) = t''(c).$$

Proof. All what holds in $\mathfrak{N}$ is forced by some p in the complete sequence $\mathcal{R}$. Hence there exists a condition $p \in \mathcal{R}$ such that

$$p \Vdash^{*} t(c,d_1) = t'(c,d_2).$$

By the restriction lemma we may assume that $\langle\langle n,i\rangle,e\rangle \in p$ implies $i \in c \cup d_1 \cup d_2$.

[more precisely, the restricted condition $p/_{c \cup d_1 \cup d_2} = p_0$ need not to be in $\mathcal{R}$, a priori, but the sequence $\mathcal{R}_0$ which starts with p_0, has p as its second element and contains then all conditions q of $\mathcal{R}$ which extend p defines obviously the same model $\mathfrak{N}$. Thus we may assume, that we have already chosen $\mathcal{R}_0$ as the sequence which defines $\mathfrak{N}$].

Define $p_0 = p_0(c) = p/_c$, $p_1 = p_1(d_1) = p/_{d_1}$ and $p_2 = p_2(d_2) = p/_{d_2}$. Hence $p_0 \cup p_1 \cup p_2 \Vdash^{*} t(c,d_1) = t'(c,d_2)$. A limited comprehension term $t''(c)$ will be found for which $p_1 \cup p_2 \cup p_3 \Vdash^{*} t(c,d_1) = t''(c)$. It will be enough to consider the case when d_2 contains only one element, say i_0. The general case follows by induction.

We shall write $i \,/\!\!/\, y$ for $\neg(i \leq j \vee j \leq i)$ and similarly $x \,\dot{/\!\!/}\, y$ for the $\mathcal{L}$-formula $\neg(x \dot{<} y \vee y \dot{<} x \vee x = y)$.

Let i_1, i_2, i_3 be elements of I, but not in $c \cup d_1 \cup \{i_0\}$, such that $i_1 < i_0 < i_3 \wedge i_2 \not\leq i_0$ and such that for $\nu = 1,2,3$ there are automorphisms τ_1, τ_2, τ_3 of $\langle I, \leq \rangle$ with $\tau_\nu(i_0) = i_\nu$ and $i \in c \cup d_1 \rightarrow \tau_\nu(i) = i$. Then by the symmetry-lemma (for $\nu = 1,2,3$):

$$p_0(c) \cup p_1(d) \cup p_2(i_\nu) \Vdash^* t(c,d_1) = t'(c,i_\nu)$$

where $\tau_\nu(p_2(i_0)) = p_2(i_\nu)$ for notational simplicity. This together with $p_0(c) \cup p_1(d) \cup p_2(i_0) \Vdash^* t(c,d_1) = t'(c,i_0)$ implies (using the restriction lemma):

$$(*) \qquad p_0(c) \cup p_2(i_0) \cup p_2(i_\nu) \Vdash^* t'(c,i_0) = t'(c,i_\nu)$$

for $\nu = 1,2,3$. We introduce the following notations. For a condition q define C_q ("the content of q") to be the following $\mathcal{L}$-sentence:

$$\bigwedge\{\underline{n} \in \dot{a}_j ;\ \langle\langle n,j\rangle,1\rangle \in q\} \wedge \bigwedge\{\neg\, \underline{m} \in \dot{a}_j ;\ \langle\langle m,j\rangle,0\rangle \in q\},$$

where $\bigwedge$ denotes conjunction. Let $C_q(^j/_x)$ be the result of replacing the generic constant $\dot{a}_j$ in C_q by the variable x.

For a finite subset s of I let D_s be the diagram of $\langle\{\dot{a}_j ; j \in s\}, \dot{<}\rangle$ i.e.:

$$D_s \triangleq \bigwedge\{\dot{a}_{j_1} \dot{<} \dot{a}_{j_2} ;\ j_1 < j_2 \wedge j_1, j_2 \in s\} \wedge \bigwedge\{\neg\, \dot{a}_{j_1} \dot{<} \dot{a}_{j_2} ;\ j_1, j_2 \in s \wedge \neg\, j_1 < j_2\}.$$

Let $D_s(^j/x)$ be the result of replacing the constant $\dot{a}_j$ in D_s by the variable x. We now claim that the following continuity-property holds:

$$(**) \quad p_0(c) \cup p_2(i_0) \Vdash^* \bigwedge_x [x \in \dot{A} \wedge C_{p_2}(^{i_0}/x) \wedge D_{c\cup\{i_0\}}(^{i_0}/x) . \rightarrow . \ t'(c,i_0) = t'(c,x)],$$

where $t'(c,x)$ results from $t'(c,i_0)$ by replacing $\dot{a}_{i_0}$ in t' by x. Let k be a limited comprehension term. We have to show that, if q is any extension of $p_0(c) \cup p_2(i_0)$ and $q \Vdash^* k \in \dot{A} \wedge C_{p_2}(^{i_0}/k) \wedge D_{c\cup\{i_0\}}(^{i_0}/k)$, then there is a $q' \supseteq q$ with $q' \Vdash^* t'(c,i_0) = t'(c,k)$.

Assume that k and q are given such that the just mentioned hypothesis is fullfilled. Since in particular $q \Vdash^* k \in \dot{A}$ we know by the definition of forcing, that there exists a $q' \supseteq q$ and a $j \in I$ such that $q' \Vdash^* k = \dot{a}_j$. Pick $\nu = 1,2$ or 3, so that $\langle\{i_0, j\}, \leq\rangle$ and $\langle i_0, i_\nu\}, \leq\rangle$ are isomorphic. We claim that there is an automorphism τ of $\langle I, \leq\rangle$ so that for this choice of ν, τ maps $\langle c \cup \{i_0, i_\nu\}, \leq\rangle$ isomorphically on $\langle c \cup \{i_0, j\}, \leq\rangle$ in such a way that τ restricted to $c \cup \{i_0\}$ is the identical mapping.

In fact, since q' extends q the first extension lemma (see page 82) yields:

$$q' \Vdash^* k = \dot{a}_j \wedge C_{p_2}(^{i_0}/_j) \wedge D_{c\cup\{i_0\}}(^{i_0}/_j),$$

which means in particular, that q' (weakly) forces that $c \cup \{i\}$ and $c \cup \{j\}$ are isomorphic. Since already $\emptyset \Vdash^* D_{c\cup\{i_0\}}(^{i_0}/_j)$ (obviously) we obtain that in fact $\langle c \cup \{i_0\}, \leqslant\rangle$ and $\langle c \cup \{j\}, \leqslant\rangle$ are isomorphic such that there is an isomorphism σ_1 leaving c pointwise fixed. But by the choice of ν there is also an isomorphism σ_2 from $\langle c \cup \{i_\nu\}, \leqslant\rangle$ onto $\langle c \cup \{i_0\}, \leqslant\rangle$ leaving c pointwise fixed. Hence $\tau_0 = \sigma_1\sigma_2$ is an isomorphism from $\langle c \cup \{i_\nu\}, \leqslant\rangle$ onto $\langle c \cup \{j\}, \leqslant\rangle$ leaving c pointwise fixed. It follows that $\langle c \cup \{i_0, j\}, \leqslant\rangle$ can be mapped isomorphically on $\langle c \cup \{i_0, i_\nu\}, \leqslant\rangle$ so that $c \cup \{i_0\}$ is left pointwise fixed. Let τ be such an isomorphism.
Since $q' \Vdash^* C_{p_2}(^{i_0}/_j)$ it must hold that $p_2(j) = \sigma_1(p_2(i_0)) \subseteq q'$.
Hence

$$p_0(c) \cup p_2(i_0) \cup p_2(j) \subseteq q'$$

and applying τ to (*) the symmetry lemma and the 1st extension lemma entail:

$$q' \Vdash^* t'(c, i_0) = t'(c, j).$$

Since $q' \Vdash^* k = \dot{a}_j$, this gives us $q' \Vdash^* t'(c, i_0) = t'(c, k)$ as desired and (**) is proved.

The limited comprehension term t"(c) can now be constructed. Suppose that $t'(c, i_0)$ is $E^\alpha y\Phi(y)$ where Φ is a formula of $\mathcal{L}$ containing the constants $\dot{a}_{i_0}$ and $\dot{a}_i$ for $i \in c$. Let $\Phi^*(x,y)$ be the $\mathcal{L}$-formula obtained from $\Phi(y)$ by replacing $\dot{a}_{i_0}$ by the variable x (at all places of occurrence) - it is assumed, that x does not occur in $\Phi(y)$.
Then define

$$t''(c) = E^\alpha y(\bigvee_x [x \in \dot{A} \wedge C_{p_2}(^{i_0}/_x) \wedge D_{c\cup\{i_0\}}(^{i_0}/_x) \wedge \Phi^*(x,y)]).$$

By (**), $p_0(c) \cup p_2(i_0) \Vdash^* t'(c, i_0) = t''(c)$. Since $p_0(c) \cup p_2(i_0) \subseteq p \in \mathfrak{G}$ and everything forced by p is true in $\mathfrak{N}$, the lemma is proved.

<u>Lemma</u>: The ordering principle (OP) holds in $\mathfrak{N}$.

<u>Proof</u>. Set up in $\mathfrak{N}$ a ramified language $\mathcal{L}^*$ with a name a* for each $a \in A$, names A^* (for A), ∇^* (for $\nabla = \mathrm{val}(\dot{<})$), names x^* for each $x \in \mathfrak{M}$, limited quantifiers $\bigvee^\alpha$, limited comprehension operators E^α (for all ordinals α of $\mathfrak{N}$) and the usual ZF-symbols. Do this in a way so that $\{\langle a, a^*\rangle ; a \in A\}$ is a set of $\mathfrak{N}$ and $\{\langle \alpha, \bigvee^\alpha\rangle ; \alpha \in On^{\mathfrak{N}}\}$ and $\{\langle \alpha, E^\alpha\rangle ; \alpha \in On^{\mathfrak{N}}\}$ are classes of $\mathfrak{N}$. Notice,

that we assume here that $\mathfrak{M}$ is $\mathfrak{N}$-definable -this assumption can be made by the remarks of page 96-97. Notice furthermore that we could not use symbols like a_i^* for $i \in I$ as names for the elements of A since the correspondence $i \mapsto a_i$ $(i \in I)$ is <u>not</u> in $\mathfrak{N}$ (for each $i \in I$ we added a generic real $a_i = \mathrm{val}(\dot{a}_i)$ to $\mathfrak{M}$, but we did <u>not</u> add the correspondence $\{\langle i,\dot{a}_i\rangle ; i \in I\}$ generically to $\mathfrak{M}$). Define an interpretation Ω^* for the constant terms of $\mathcal{L}^*$ inductively by setting:

$$\begin{aligned} &\Omega^*(a^*) = a && \text{for each } a \in A,\\ &\Omega^*(A^*) = A &&\\ &\Omega^*(\nabla^*) = \nabla , \text{ and} &&\\ &\Omega^*(x^*) = x && \text{for each } x \in \mathfrak{M}, \end{aligned}$$

and then extending to all limited comprehension terms of $\mathcal{L}^*$, so that Ω^* is $\mathfrak{N}$-definable. Now define $\mathrm{supp}^*(u)$, for $u \in \mathfrak{N}$, as the finite subset, call it d, of A of minimal cardinality such that there is a term $t^*(d)$ of $\mathcal{L}^*$ mentioning only the names a^* for $a \in d$ with $\Omega^*(t^*(d)) = u$. Read $\mathrm{supp}^*(u)$ as: the support of u. By the previous lemma, supp^* is always well defined. Notice, that there is a clear one-to-one-correspondence * between constant terms t of $\mathcal{L}$ and terms t^* of $\mathcal{L}^*$ so that $\Omega^*(t^*) = \mathrm{val}(\bar{t})$. Further, supp^* is $\mathfrak{N}$-definable. For each finite subset d of A let

$$V_d = \{x \in \mathfrak{N} ; \mathrm{supp}^*(x) = d\}$$

Each class V_d has an $\mathfrak{N}$-definable wellordering: the symbols from the alphabet of $\mathcal{L}^*$ used to construct constant terms t^* with $\mathrm{supp}^*(\Omega^*(t^*)) = d$ have an $\mathfrak{N}$-definable wellordering and thus the constant terms t^* of $\mathcal{L}^*$ with $\mathrm{supp}^*(\Omega^*(t^*)) = d$ have an $\mathfrak{N}$-definable wellordering (e.g. the lexicographic ordering as modified by Gödel [25] p.36). Using the interpretation Ω^* one obtains an induced well-ordering of V_d. Each d is a finite subset of A, which is linearly ordered, being a subset of the real line. The set of finite subsets d of A can be linearly ordered e.g. by the usual lexicographic method. [For more details see the proof of the theorem on page 97-98].

Hence let <u>Lex</u> be the lexicographic ordering of finite subsets of A and for each finite subset d of A let W_d be the wellordering of V_d. Now if u is any set of $\mathfrak{N}$, then

$$\{\langle x_1,x_2\rangle ; x_1,x_2 \in u \wedge [\langle \mathrm{supp}^*(x_1),\mathrm{supp}^*(x_2)\rangle \in \underline{\mathrm{Lex}} \vee \vee (\mathrm{supp}^*(x_1) = d = \mathrm{supp}^*(x_2) \wedge \langle x_1,x_2\rangle \in W_d)]\}$$

is a linear ordering of u. This proves the lemma and hence the theorem of A.R.D. Mathias is proved.

Let (AC_{wo}) (in the notation of A.Lévy [39]p.223) be the following consequence of the usual axiom of choice:

(AC_{wo}): $\bigwedge_x [\bigwedge_y (y \in x \to y \neq \emptyset \wedge y$ can be wellordered) $\to$
$\bigvee_f (Fnc(f) \wedge Dom(f) = x \wedge \bigwedge_{y \in x} f(y) \in y)]$.

Theorem (A.R.D. Mathias): If ZF is consistent, then ZF + (OP) + (AC_{wo}) + $\neg$ (AC) is consistent too. Thus the Axiom of choice (AC) is independent from (AC_{wo}) + Ordering principle (OP) in ZF.

Proof ([58]): Take the model $\mathfrak{N}$ constructed above. A set z is in $\mathfrak{N}$ wellorderable iff there is a finite subset d of A such that for all y, if $y \in z$, then $supp^{*}(y) \subseteq d$. Hence, if z is wellorderable in $\mathfrak{N}$, then the $\mathfrak{N}$-definable linear-ordering turns out to be a well-ordering. Thus, if x is a set o well-orderable sets, then there is a function f in $\mathfrak{N}$ which assigns to each $z \in x$ a wellordering. Hence (AC_{wo}) holds in $\mathfrak{N}$. As it was shown in preceeding lemmata, (OP) and $\neg$ (AC) are true in $\mathfrak{N}$. This proves the theorem.

F) THE KINNA-WAGNER CHOICE PRINCIPLE

An interesting weakened version of the axiom of choice has been considered by W.Kinna and K.Wagner in their paper:

[44] W.KINNA-K.WAGNER: Ueber eine Abschwächung des Auswahlpostulates; Fund. Math. 42(1955)p.75-82.

In contrast to the usual (AC) where the choice function selects one element, the functions considered by Kinna-Wagner select non-empty, but proper subsets:

(KW-AC): The Kinna-Wagner choice principle: If x is a set all of whose elements have at least two elements, then there exists a function f, defined on x, such that for all $y \in x$, $\emptyset \neq f(y) \subsetneq y$.

$$\bigwedge_x [\bigwedge_y (y \in x \to 2 \leq \overline{\overline{y}}) \to \bigvee_f (Fnc(f) \wedge Dom(f) = x \wedge \bigwedge_{y \in x} (\emptyset \neq f(y) \subseteq y \wedge y \neq f(y)))].$$

Obviously (KW-AC) is a consequence of (AC). W.Kinna and K.Wagner have shown, that (KW-AC) implies the ordering principle (OP) "Every set can be totally ordered":

Lemma (Kinna-Wagner[44]): ZF $\vdash$ (KW-AC) $\rightarrow$ (OP).

Sketch of Proof. Let M be a given set and let f be a function, defined on P(M), the power set of M, such that $\emptyset \neq f(y) \neq y$, $f(y) \subset y$ whenever $y \in P(M)$ and y contains at least two elements. Following the $\bigcap$-method of Zermelo's second proof for (AC) $\rightarrow$"Well-ordering-theorem" (see: Math. Amm. 65 (1908)p.107-128) one starts the proof with the following definition:

Let F be the least family of subsets of M satisfying:

(1) $M \in F$,

(2) $(x \in F \wedge 2 \leqslant \bar{\bar{x}}) \rightarrow (f(x) \in F \wedge x - f(x) \in F)$,

(3) $t \subseteq F \rightarrow \bigcap t \in F$.

Define an element e of F to be normal iff the following holds:

$\bigwedge_{x \in F}[e \subseteq x \wedge e \neq x \rightarrow e \subseteq f(x) \vee e \subseteq x - f(x)]$.

There are normal elements in F, e.g. M is normal. For e normal, define:

$g_e = \{x \in F;\ x \cap e = \emptyset \vee [x \cap e \neq \emptyset \rightarrow (e \subseteq x \vee x \subseteq f(e) \vee x \subseteq e - f(e)]\}$.

One proves that g_e satisfies the conditions (1), (2), (3) and is therefore (by the minimality of F) equal to F. This shows that if e is normal, then

(*) $\bigwedge_{x \in F}[x \cap e = \emptyset \vee (x \cap e \neq \emptyset \wedge [e \subseteq x \vee x \subseteq f(e) \vee x \subseteq e - f(e)])]$

holds. Now, let N be the set of normal elements of F, then N satisfies conditions (1), (2), (3) and thus N = F by the minimality of F. This shows that every element of F is normal. Intuitively this means the following: by means of an iterated application of f to M, then to f(M) and M-f(M), then to f(f(M)), f(M)-f(f(M)), f(M-f(M)), (M-f(M)) - f(M-f(M)), etc.... one obtains by transfinite induction the following binary tree:

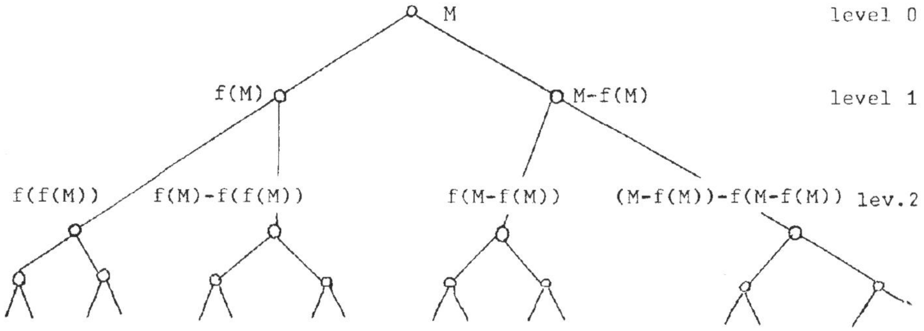

where at each point x the tree splits into two branches, if $\overline{\overline{x}} \geqslant 2$, where f(x) is the successor of x at the left branch and x - f(x) the successor of x on the right branch. The information, that every element is normal means that if $e \in F$ and $x \in F$, then either e and x are on different branches, or, if there is a branch going through x and e then either the level of x is smaller than the level of e (case: $e \subseteq x$) or x appears on the branch after one of the two successors of e. With this in mind, it is not difficult to prove, that the binary relation ρ, defined on M by:

$$p \,\rho\, q \leftrightarrow p \in f(\bigcap\{x \in F;\ p,q \in x\} \ \vee\ q \notin f(\bigcap\{x \in F;\ p,q \in x\}).$$

is a linear ordering of M. This proves the lemma.

The proof shows, that a binary, wellfounded tree can be embedded into P(M), the powerset of M, such that the image forms a chain C in P(M). The tree itself is equipotent with some ordinal α, hence $\overline{\overline{P(M)}} \geqslant \overline{\overline{\alpha}}$. On the other hand, if D(C) is the Dedekind completion of C, then D(C) is a maximal chain in P(M), linking $\emptyset$ and M. Hence $\overline{\overline{M}} \leqslant 2^{\alpha}$. This proves the following:

<u>Theorem</u> (Kinna-Wagner [44]) In ZF the statement (KW-AC) is equivalent to the statement:

(KW-O): For every set M there is an ordinal α such that there exists a one-to-one function mapping M into $P(\alpha)$, the powerset of α.

(KW-O) is a strong form of the Ordering-principle, since it asserts that every set M has a linear ordering ρ which is a subset of the canonical ordering $\leqslant$ on $P(\alpha)$, for some ordinal α, defined by $x \leqslant y \leftrightarrow \mathrm{Min}\{(x \cup y) - (x \cap y)\} \in y$. Since ZF $\vdash$ (AC) $\rightarrow$ (KW-AC) $\leftrightarrow$ (KW-O) $\rightarrow$ (OP) we ask, whether the first or the last arrow can be reversed. It is known, that both cannot be reversed. J.D.Halpern and A.Lévy have first shown, that (AC) does not follow from (KW-AC) in ZF (see [35], and:

[34] J.D.HALPERN-A.LÉVY: the Ordering Theorem does not imply the axiom of choice; Notices of the Amer. Math. Soc. 11 (January 1964)p.56.

The problem, whether (OP) $\rightarrow$ (KW-O) can be proved in ZF remained for some years open, though Mostowki has shown already in 1958, that in ZF^0 (id est ZF but without the axiom of foundation) (OP) does not imply (KW-O), see:

[67] A. MOSTOWSKI: On a problem of W.Kinna and K.Wagner; Coll. Math. vol.6(1958)p.207-208.

The independence of (KW-O) from (OP) in ZF^0 follows more directly already from the fact, that (PW) holds in every Fraenkel-Mostowski-Specker model (see p.62), hence in particular in Mostowski's model of [64] in which (OP) holds, while (AC) fails. If (KW-O) would hold there, then (AC) would also hold. In 1969 U.Felgner has shown, that also in full ZF, (KW-O) is independent from (OP).

[18] U.FELGNER: Das Ordnungstheorem impliziert nicht das Kinna-Wagnersche Auswahlprinzip. To appear.

Here we shall sketch the proofs for both results. But, in order to prove that (AC) is independent from (KW-AC) we shall not use the model $\mathfrak{M}[a_0, a_1, \ldots, A]$ (a_n generic reals for $n \in \omega$, $A = \{a_n; n \in \omega\}$) which was used originally by Halpern and Lévy (for a more detailed description of this model, see sections C and G). We shall use the model of Mathias, described in section E, since this will give us the additional information, that (KW-AC) does not imply the order-extension principle (OE).

Theorem. The choice principle (KW-AC) of Kinna and Wagner holds in the model of Mathias. Hence, if ZF is consistent, then (KW-AC) does not imply the orderextension principle (OE), though (KW-AC) implies the ordering principle (OP).

Proof. We have shown in section E, that Mathias' model $\mathfrak{N}$ has the following features: for every set u of $\mathfrak{N}$, the following relation is a linear ordering of u:

$$\{(x_1, x_2);\ x_1, x_2 \in u \wedge [(\mathrm{supp}^*(x_1), \mathrm{supp}^*(x_2)) \in \underline{Lex} \vee (\mathrm{supp}^*(x_1) = \mathrm{supp}^*(x_2) = d \wedge (x_1, x_2) \in W_d)]\}$$

Here Lex was the lexicographic ordering of finite subsets of A, the set of generic reals a_i, and W_d was the wellordering of $V_d = \{x; \mathrm{supp}^*(x) = d\}$. Since $A \subseteq 2^{\aleph_0}$, every set u of $\mathfrak{N}$ splits into $\leq 2^{\aleph_0}$ many wellordered subsets. Thus we can compute an ordinal λ such that u can be embedded (i.e. mapped into, by a one-to-one function) into $\aleph_\lambda \times 2^{\aleph_0}$. Thus u can be embedded into $2^{\aleph_\alpha}$, the powerset of $\aleph_\alpha$, for a sufficient large ordinal α. Thus (KW-O) holds in $\mathfrak{N}$. The rest follows from results, proved in section E, q.e.d. (take e.g. $\alpha = \lambda$)

Theorem. (U.Felgner): If ZF is consistent, then (KW-O) is independent from the ordering principle (OP).

Sketch of proof. Let $\mathfrak{M}$ be a countable standard model of ZF + V = L. Let Q be the set in $\mathfrak{M}$ of rational numbers and let $\leqslant$ be the usual ordering (of type η_0) of Q. Define in $\mathfrak{M}$ a ramified language $\mathcal{L}$ which contains besides the usual ZF-symbols and the useful limited quantifiers $\bigvee^{\alpha}$ and limited comprehension terms E^{α} $(\alpha \in On^{\mathfrak{M}})$, constants $\dot{a}_{i,j}$, $\dot{b}_i$, for $i \in Q$ and $j \in \omega$, a constant $\dot{c}$ and a binary predicate $\dot{\leqslant}$. Let $\mathcal{L}$ further contain constants $\underline{x}$ for each $x \in \mathfrak{M}$ and a unary predicate $\dot{g}$, which will be used to make $\mathfrak{M}$ an $\mathfrak{N}$-definable class.

A condition p is a finite partial function from $\omega \times Q \times \omega$ into 2. A forcing relation $\Vdash$ is defined as usual containing the following key-clauses (here let t be any term of $\mathcal{L}$):

$$p \Vdash t \in \dot{a}_{i,j} \Leftrightarrow (\exists n \in \omega)(p \Vdash \underline{n} \simeq t \;\&\; p(\langle n,i,j\rangle) = 1);$$

$$p \Vdash t \in \dot{b}_i \Leftrightarrow (\exists j \in \omega)(p \Vdash t \simeq \dot{a}_{i,j});$$

$$p \Vdash t \in \dot{c} \Leftrightarrow (\exists i \in Q)(p \Vdash t \simeq \dot{b}_i);$$

$$p \Vdash t_1 \dot{\leqslant} t_2 \Leftrightarrow (\exists i_1, i_2 \in Q)(i_1 \leqslant i_2 \;\&\; p \Vdash t_1 \simeq \dot{b}_{i_1} \;\&\; p \Vdash t_2 \simeq \dot{b}_{i_2}).$$

Obtain a complete sequence $\mathcal{R}$ of conditions and thereby a valuation val of the constant terms of $\mathcal{L}$, which gives the model $\mathfrak{N}$. It holds that $a_{i,j} = val(\dot{a}_{i,j}) \subseteq \omega$, $b_i = val(\dot{b}_i) = \{\dot{a}_{i,j};\, j \in \omega\}$, $c = val(\dot{c}) = \{b_i;\, i \in Q\}$ and $val(E^{\omega+3}x(\bigvee_{y_1}\bigvee_{y_2}(x = \langle y_1,y_2\rangle \wedge y_1 \dot{\leqslant} y_2))) = \trianglelefteq$ is a linear ordering of c. Notice, that inside of $\mathfrak{N}$ there is no isomorphism between $\langle Q,\leqslant\rangle$ and $\langle c,\trianglelefteq\rangle$, but outside of $\mathfrak{N}$ both are isomorphic.

Symmetries. Let $\mathfrak{G}$ be the group in $\mathfrak{M}$ of all orderpreserving one-to-one mappings from Q onto Q and let $\mathcal{H}$ be in $\mathfrak{M}$ the group of all permutations of ω, which move only finitely many elements of ω.

(1) $\mathfrak{G}$ operates on $\mathcal{L}$ by setting: if $\pi \in \mathfrak{G}$, then $\pi(\dot{a}_{i,j}) = \dot{a}_{\pi(i),j}$, $\pi(\dot{b}_i) = \dot{b}_{\pi(i)}$; π acts as the identity on all other symbols of $\mathcal{L}$.

(2) $\mathcal{H}$ operates on $\mathcal{L}$ by setting: if $\langle\tau,i\rangle \in \mathcal{H} \times Q$, then define $\tau_i = \langle\tau,i\rangle$ and then: $\tau_i(\dot{a}_{i,j}) = \dot{a}_{i,\tau(j)}$, $\tau_i(\dot{a}_{k,j}) = \dot{a}_{k,j}$ if $k \neq i$, τ_i acts identical on all other symbols of $\mathcal{L}$.

(3) If p is a condition, $\pi \in \mathfrak{G}$, $\tau_i = \langle\tau,i\rangle \in \mathcal{H} \times Q$, then define:
$\pi(p) = \{\langle\langle n,\pi(i),j\rangle,e\rangle;\ \langle\langle n,i,j\rangle,e\rangle \in p\}$
$\tau_i(p) = \{\langle\langle n,i,\tau(j)\rangle,e\rangle;\langle\langle n,i,j\rangle,e\rangle \in p\} \cup \{\langle\langle n,k,j\rangle,e\rangle \in p;\ k \neq i\}$.

We have two symmetry- lemmata:

First symmetry-lemma: If $\pi \in \mathfrak{G}$, then $p \Vdash \Phi \leftrightarrow \pi(p) \Vdash \pi(\Phi)$.

Second symmetry-lemma: If $\tau_i \in \mathcal{H} \times Q$, then $p \Vdash \Phi \Leftrightarrow \tau_i(p) \Vdash \tau_i(\Phi)$.

Restriction lemma: If $p \Vdash \Phi$ and $p_0 = \{\langle\langle m,i,j\rangle ,e\rangle \in p;\ \dot{a}_{i,j}$ occurs in $\Phi\}$, then $p_0 \Vdash^* \Phi$.

For a $\mathcal{L}$-formula Φ define $occ_1(\Phi)$ to be the set of ordered pairs $\langle i,j\rangle \in Q \times \omega$ such that $\dot{a}_{i,j}$ occurs in Φ, and let $occ_2(\Phi)$ be the set of rationals i such that $\dot{b}_i$ occurs in Φ. For a limited comprehension term $t = E^\alpha \times \Phi$ write $t = t(\Delta_1,\Delta_2)$ in order to indicate that $\Delta_1 = occ_1(\Phi)$ and $\Delta_2 = occ_2(\Phi)$. Further, let $pr_1(\Delta_1) = \{i \in Q; \bigvee_{j\in\omega} \langle u,j\rangle \in \Delta_1\}$ be the projection of Δ_1 to the first coordinate . The following support-lemma is a generalization of the corresponding support-lemmata of Mostowski [64] and Mathias [58].

Support lemma. Let $t(\Delta_1,\Delta_2)$ and $t^*(\Delta_1^*,\Delta_2^*)$ be limited comprehension terms of $\mathcal{L}$ such that

$$\mathfrak{N} \models t(\Delta_1,\Delta_2) = t^*(\Delta_1^*,\Delta_2^*);$$

then there exists a limited comprehension term $t^0 = t^0(\Delta_1^0,\Delta_2^0)$ such that $\Delta_1^0 = \Delta_1 \cap \Delta_1^*$ and $\Delta_2^0 = (\Delta_2 \cap \Delta_2^*) \cup pr_1(\Delta_1^0)$, and $\mathfrak{N} \models t(\Delta_1,\Delta_2) = t^0(\Delta_1^0,\Delta_2^0)$.

The proof is along the lines of the support lemma proved in section E, but in the present case slightly more complicated. Now, it can be verified, that the ordering principle (OP) holds in $\mathfrak{N}$: define $V_{\langle\Delta_1,\Delta_2\rangle} = \{x;\ supp_1^*(x) = \Delta_1 \wedge supp_2(x) = \Delta_2\}$, where $supp_1^*(u)$, and $supp_2^*(u)$ are the finite sets such that $supp_1^*(u)$ is of minimal cardinality, $supp_2^*(u)$ is of minimal cardinality modulo $supp_1^*(u)$, such that there is a term $t = E^\alpha \times \Phi(x)$ with $occ_1(\Phi) = supp_1^*(u)$, $occ_2(\Phi) = supp_2^*(u)$ and t is interpreted by u. Each class $V_{\langle\Delta_1,\Delta_2\rangle}$ has a definable wellordering. This together with a linear ordering of the ordered pairs $\langle\Delta_1,\Delta_2\rangle$ can be used to obtain total orderings for any set of $\mathfrak{N}$ (see the corresponding proof in section E).

In a next lemma one shows that in $\mathfrak{N}$ every subset of $c = val(\dot{c})$ is a finite union of open, closed or at one side open, at the other side closed intervals from $\langle c,\trianglelefteq\rangle$ (use the 2nd-symmetry-lemma and the restriction lemma to see this). It follows, that in $\mathfrak{N}$ every subset of c is definable by a formula $\Phi(x)$ of $\mathcal{L}$ in which none of the symbols $\dot{a}_{i,j}$ occurs, since only $\dot{\leq}$ and names $\dot{b}_i$ for the endpoints of the intervals are used. Now the argument of Mostowski [67] can be used to show, that (KW-AC) fails in $\mathfrak{N}$ by showing, that there is no function f in $\mathfrak{N}$ selecting from each proper, closed interval of

$(c, \trianglelefteq)$ a non-empty, proper subset. The arguments given above show, that in a transscription of Mostowski's continuity argument, by the restriction lemma, the forcing conditions do not pose additional problems. This proves the theorem.

Notice, that in contrast to all our examples of generic extensions, here we have used a somehow different approach. We did not add a Cohen generic copy of η_0 as a subset of $2^{\aleph_0}$ to the groundmodel (this would yield a model, in which (KW-AC) holds), but a copy of η_0 as a subset of the powerset of $2^{\aleph_0}$. Thus we have shown that neither (KW-AC) $\rightarrow$ (AC) nor (OP) $\rightarrow$ (KW-AC) is provable in **ZF**.

G) THE INDEPENDENCE OF THE AXIOM OF CHOICE (AC) FROM THE BOOLEAN PRIME IDEAL THEOREM (BPI)

In section E we considered the Boolean Prime Ideal theorem (BPI) and noticed that in ZF the (BPI) is a consequence of (AC). Here we shall prove, that the converse is not true, namely that (BPI) $\rightarrow$ (AC) is not a theorem of ZF. As we mentioned proviously, this result is due to J.D.Halpern - A.Lévy **[35]**. A short outline of this prove is contained in:

[31] J.D.HALPERN: The Boolean Prime Ideal Theorem; Lecture Notes prepared in connection with the Summer Institute on Axiomatic Set Theory at UCLA, July 10 - August 4, 1967 (informally distributed manuscripts), 7 pages.

Historically Halpern first showed in 1962 in his doctoral dissertation that in the model $\mathfrak{M}$ of Mostowski **[64]** the (BPI) holds, thus proving that in ZF^0 (i.e. ZF without foundation) the (BPI) does not imply (AC). Mostowski had shown in 1939 in **[64]** that in $\mathfrak{M}$ the ordering principle (OP) holds while (AC) is violated in $\mathfrak{M}$. Halpern's result appeared in print:

[32] J.D.HALPERN: The independence of the axiom of choice from the Boolean prime ideal theorem; Fund. Math. 55(1964) p.57-66.

After Cohen's invention of the generic ZF-model's in 1963, Halpern proved in collaboration with A-Lévy and H.Läuchli (via generic models), that also in full ZF the (AC) is independent from the (BPI) - see **[35]** and **[31]**.

According to the tradition of our lecture notes we shall start to explain the ideas behind the construction of a model of set theory in which (BPI) + $\neg$ (AC) holds by discussing Halpern's original proof [32], that in Mostowski's model $\mathfrak{M}$ (BPI) holds. But first of all we need two lemmata.

Lemma 1([32] p.62): Let $\mathbb{B} = \langle B, \sqcup, - \rangle$ be a Boolean algebra, Aut($\mathbb{B}$) be the group of all automorphisms of $\mathbb{B}$ and let H be any subgroup of Aut($\mathbb{B}$). If I is an ideal of $\mathbb{B}$, closed under H, and if $b \in B$ and J is the smallest ideal closed under H which includes I and $\{b\}$ and if $1_B \in J$, then there is a finite subset S of H such that

$$\bigsqcap_{\phi \in S}(\phi(-b)) \in I.$$

Proof. $\bigsqcap\{\phi(-b);\ \phi \in S\}$ is the greatest lower bound of the set of elements $\phi(-b)$ of B for $\phi \in S$. Let I(b) be the ideal generated by I and $\{b\}$; then $I(b) = \{x \in B; (\exists y \in I)(x \leqslant y \sqcup b)\}$. Close I(b) under automorphisms ϕ of H: $J_0 = \{z \in B;\ (\exists y \in I(b))(\exists \phi \in H)[z \leqslant \phi(y)]\}$ and let J be the ideal generated by J_0, id est: $J = \{\bigsqcup e;\ e$ is a finite subset of $J_0\}$. By assumption $1_B \in J$, hence there is a finite subset $e = \{x_1, \dots, x_n\}$ of J_0 such that $1 = \bigsqcup e$. Since $x_i \in J_0$, there are $y_i \in I$ and $\phi_i \in H$ $(i = 1,2,\dots,n)$ such that $x_i \leqslant \phi_i(y_i \sqcup b)$. Hence

$$1_B = \bigsqcup_{i=1}^{n} x_i = \bigsqcup_{i=1}^{n}\phi_i(y_i \sqcup b) = \bigsqcup_{i=1}^{n}\phi_i(y_i) \sqcup \bigsqcup_{i=1}^{n}\phi_i(b)$$

and hence by taking the complements:

$$-(\bigsqcup_{i=1}^{n}\phi_i(y_i) \sqcup \bigsqcup_{i=1}^{n}\phi_i(b)) = \bigsqcap_{i=1}^{n}\phi_i(-y_i) \sqcap \bigsqcap_{i=1}^{n}\phi_i(-b) = 0_B \in I$$

Since I is closed under H, it follows that $\bigsqcup\{\phi_i(y_i);\ i = 1,\dots,n\} \in I$. Since $u \sqcap v = o$ implies $v \leqslant -u$ in a boolean algebra, put $u = \bigsqcap\{\phi_i(-y_i);\ 1 \leqslant i \leqslant n\}$ and $v = \bigsqcap\{\phi_i(-b);\ 1 \leqslant i \leqslant n\}$, then it follows $v \leqslant -u \in I$, hence $v \in I$ and if we define $S = \{\phi_i;\ 1 \leqslant i \leqslant n\}$, then the lemma is proved.

Lemma 2 ([32] p.62). If $\mathbb{B}$ is a Boolean algebra, X a finite subset of B and P is the set of all functions f on X such that $f(x) \in \{x,-x\}$, then

$$\bigsqcup_{f \in P}(\bigsqcap_{x \in X} f(x)) = 1_B$$

Proof by induction on the cardinality of X.

The (BPI) holds in Mostowski's model $\mathfrak{M}$. Take countably many atoms (urelemants of reflexive sets $x = \{x\}$) ordered of type η_0. Let G be the group of orderpreserving mappings and F be the filter of subgroups generated by the finite support subgroups. Mostowski showed in [64], that in $\mathfrak{M} = \mathfrak{M}[G,F]$, (OP) holds while (AC) fails (see section E of this chapter and chapter III for details and notation). J.D.Halpern extended in [32] this result by showing that in $\mathfrak{M}$ even the (BPI) holds. He proceeds as follows. Let $\mathbb{B}$ be a boolean algebra in $\mathfrak{M}$. Then $H[\mathbb{B}] = \{\phi \in G;\ \phi(\mathbb{B}) = \mathbb{B}\} \in F$ and by the definition of F there is a finite support subgroup $K[e] = \{\phi \in G;\ \phi$ is identical on $e\}$ (e a finite set of atoms) such that $K[e] \leqslant H[\mathbb{B}]$. Thus $\mathbb{B}$ is e-symmetric and every $\phi \in K[e]$ is an automorphism of $\mathbb{B}$. Halpern shows, that among the e-symmetric ideals of B there is a prime ideal I. In fact consider (outside of $\mathfrak{M}$) the set

$$Z = \{J;\ J \text{ is an ideal of } \mathbb{B} \wedge J \in \mathfrak{M} \wedge 1_B \notin J \wedge K[e] \leqslant H[J]\}.$$

Z is inductively ordered by $\subseteq$ (see the proof of a similar situation on p.65), and has hence by Zorn-Kuratowski's lemma a maximal element, say I_0. We claim, that I_0 is a prime ideal of B (I_0 is e-symmetric). Suppose not, then there exists $b \in B$ such that $b \notin I_0$ and $(-b) \notin I_0$. Let I_1 be the smallest ideal of B which includes I_0 and b and is closed under $K[e]$, and let I_2 be the smallest ideal of B which includes I_0 and $-b$ and is closed under $K[e]$. Since both, I_1 and I_2, are e-symmetric ideals, hence in $\mathfrak{M}$, they cannot satisfy the hypothesis not to include 1_B, since otherwise $I_1 \in Z$ and $I_2 \in Z$, contradicting the maximality of I_0. Hence $1_B \in I_1$ and $1_B \in I_2$ and by lemma 1 there are finite subsets S_1, S_2 of $K[e]$, such that

(1) $$\prod \{\phi(-b);\ \phi \in S_1\} \in I_0$$

and

(2) $$\prod \{\phi(b);\ \phi \in S_2\} \in I_0 .$$

Let $r = \overline{\overline{e}}$, then r determines r + 1 open intervals K_i ($0 \leqslant i \leqslant r$) of A (in the ordering of type of the rationals). We want to get (via lemma 2) the contradiction that $1_B \in I_0$. To do so, we need a certain finite subset X of B such that $\prod \{f(x);\ x \in X\} \in I_0$ for all fuctions f on X such that $f(x) \in \{x,-x\}$; then lemma 2 entails that $1_B \in I_0$. Since one wan'ts to derive $\prod \{f(x);\ x \in X\} \in I_0$ by some permutation arguments from (1) and (2), Halpern finds (using a combinatorial theorem of R.Rado) a certain finite subset W of the set A of atoms and takes then as $X = \{x \in B;\ \exists \phi \in K[e] : \phi^*(b) = x \wedge \phi(g) - e \subseteq W\}$, where ϕ^* is the unique extension of ϕ to an automorphism of the whole universe (see p.53), and g is the support of $b \in B$. Rado's theorem gives only in dependence of $\overline{\overline{S_1}}$, $\overline{\overline{S_2}}$ and $\overline{\overline{K_i \cap g}}$ ($i \in r + 1$) a certain finite cardinal number q. The property, that the atoms are totally ordered of type η_0 is essentially used to conclude, that between all points of **W** we can embed q points. W has a certain partition property which is used to find an automorphism ψ of B such that either all elements of $\{\phi(b);\ \phi \in S_2\}$ or $\{\phi(-b);\ \phi \in S_1\}$ can be mapped into $\{f(x);\ x \in X\}$. Then $\prod\{f(x);\ x \in X\} \leq \psi(\prod\{\phi(-b);\ \phi \in S_1\})$ or $\prod\{f(x);\ x \in X\} \leq \psi(\prod\{\phi(b);\ \phi \in S_2\})$ for $\psi \in K[e]$. Since I_0 is closed under $K[e]$, it follows from (1) and (2) that $\psi(\prod\{\phi(-b);\ \phi \in S_1\}) \in I_0$ and $\psi(\prod\{\phi(b);\ \phi \in S_2\}) \in I_0$. Hence, since I_0 is an ideal, $\prod\{f(x);\ x \in X\} \in I_0$ for all f under consideration. As indicated above, this yields $1_B \in I_0$, a contradiction! Hence I_0 is prime. For all details of the proof, the reader must be referred to Halpern's paper [32].

It is possible to carry over these ideas to the construction of a Cohen-generic model $\mathfrak{N}$ of ZF + (BPI) + $\neg$ (AC). This has been done in collaboration by Halpern, Lévy and Läuchli. The construction of the model has been carried out by Halpern-Lévy [35]. In their construction a combinatorial argument was used (different from Rado's theorem), which has been established by Halpern and Läuchli in:

[33] J.D.HALPERN - H.LÄUCHLI: A partition theorem; Transactions Amer. Math. Soc. vol.124(1966)p.360-367.

The model constructed by Halpern-Lévy is a boolean valued model. We shall, however, construct a Cohen-generic extension by means of forcing. Our remark, that in the case of permutation models the atoms have to be linearly ordered of type η_0, suggests that in the case of Cohen generic models $\mathfrak{N}$ the generic reals $\dot{a}_i$ (for $i \in \omega$) have to form a dense subset of the real line of $\mathfrak{N}$. Following this idea we construct $\mathfrak{N}$ by adding to some countable standard model $\mathfrak{M}$ infinitely many Cohen-generic reals $\dot{a}_i$ ($i \in \omega$) and generically a set A which just collects these reals $\dot{a}_i$. We shall use the notation $\mathfrak{N} \triangleq \mathfrak{M}[a_0, a_1, \ldots, A]$. This model $\mathfrak{N}$ has been described in section C of this chapter on pages 101 - 103.

<u>Theorem</u> (Halpern-Lévy): $\mathfrak{M}[a_0, a_1, \ldots, A]$ is a model of ZF + (BPI) + $\neg$ (AC). Hence, if ZF is consistent, then (BPI) $\rightarrow$ (AC) is not provable in ZF.

<u>Proof</u>. A detailed presentation of $\mathfrak{M}[a_0, a_1, \ldots, A]$ has been given in section C, where it was shown, that in this model A is an infinite, but Dedekind-finite set. Hence (AC) does not hold in hold in this

model. The proof, that (BPI) holds will require several lemmata.

First we remind our reader to the following (see p.99 and 102):

<u>Symmetry-lemma</u>: Let G be in $\mathfrak{M}$ the group of all permutations of ω. for any $\sigma \in G$, any condition $p: \omega \times \omega \rightarrow 2$ and any $\mathcal{L}$-sentence Φ, $p \Vdash \Phi \Leftrightarrow \sigma(p) \Vdash \sigma(\Phi)$.

<u>Restriction Lemma</u>. Let p be a condition, Φ a sentence of $\mathcal{L}$, $occ(\Phi)$ the finite set of natural numbers i such that $\dot{a}_i$ occurs in Φ, and let $p_0 = p/_{occ(\Phi)}$ be $\{\langle\langle n,i\rangle ,e\rangle \in p; i \in occ(\Phi)\}$. If $p \Vdash \Phi$, then $p_0 \Vdash^* \Phi$.

<u>Proof</u>. Suppose $p \Vdash \Phi$ and $\sim p_0 \Vdash^* \Phi$. Then there is a condition $q \supseteq p_0$, such that $q \Vdash \neg \Phi$. Define $c_0 = occ(\Phi)$,

$$d_1 = \{i \in \omega; \bigvee_{n\in\omega} \bigvee_{e\in 2}(\langle\langle n,i\rangle ,e\rangle \in p \wedge i \notin c_0)\}$$

$$d_2 = \{i \in \omega; \bigvee_{n\in\omega} \bigvee_{e\in 2}(\langle\langle n,i\rangle ,e\rangle \in q \wedge i \notin c_0)\}$$

Now let σ be any permutation of ω which leaves c_0 pointwise fixed, and maps d_1 in $\omega-(c_0 \cup d_2)$. Then $Dom(q) \cap Dom(\sigma(p)) \subseteq \omega \times c_0$. Since both, q and $\sigma(p)$, extend p_0, they coincide on the common part of their domain. Hence $\sigma(p) \cup q$ is a function, hence also a condition. Since σ leaves c_0 pointwise fixed: $\sigma(\Phi) = \Phi$. Thus the symmetry lemma tells us, that $\sigma(p) \Vdash \Phi$. Together with $q \Vdash \neg \Phi$, the extension lemma yields $\sigma(p) \cup q \Vdash \Phi \wedge \neg \Phi$, a contradiction. Hence $p_0 \Vdash^* \Phi$ must be true.

The next lemma will say, that $A \triangleq \{a_i; i \in \omega\}$ is a dense subset of 2^ω in the product topology of 2^ω. The following relation $<$ is the usual linear ordering of 2^ω:

$$s_1 < s_2 \Leftrightarrow Min((s_1 - s_2) \cup (s_2 - s_1)) \in s_2$$

Since 2^ω is considered as the product of $\aleph_0$ copies of the two-point discrete space $2 = \{0,1\}$, we may endow 2^ω with the product topology, i.e. the <u>basic</u> open sets are of the form

$$b_r = \{f \in 2^\omega; f \supseteq r\}$$

where r is a finite partial function from ω into 2. Hence b_r is the set of functions (=real numbers) from ω into 2 which extend r. The space 2^ω endowed with this topology is called the Cantor-space, since 2^ω is homeomorphic to the Cantor-discontinuum (considered as a subspace of the real line with the usual interval-topology) - see e.g. Ph.Dwinger: Introduction to Boolean Algebras, Physica-Verlag Würzburg 1961, p.49-50, or R.Sikorski: Boolean algebras, Springer Verlag Berlin 1964, p.43, and textbooks on Topology, e.g. Kelley. The space 2^ω with the topology as given above is a totally disconnected compact Hausdorff-space and the basic open sets b_r are also closed, and hence regular open sets.

Lemma 3. Every basic open set b_r of 2^ω contains a generic real a_i, or better stated: $\mathfrak{N} \models$ "$b_r \cap A \neq \emptyset$ for every finite partial function $r : \omega \to 2$".
Hence it holds in $\mathfrak{N}$ that A is a dense subset of 2^ω.

Proof. suppose $b_r \cap A \neq \emptyset$ does not hold in $\mathfrak{N}$ for every basic open set b_r. Hence, there is a finite partial function $r : \omega \to 2$ in such that $b_r \cap A = \emptyset$ holds in $\mathfrak{N}$. Since everything which holds in $\mathfrak{N}$ is forced by some condition in the complete sequence $\mathcal{R}$ (which defined $\mathfrak{N}$), there is $p \in \mathcal{R}$ such that

$$p \Vdash b_{\underline{r}} \cap A = \emptyset$$

Since p is finite, there is a natural number i_0 such that for every $n \in \omega$, $\langle n,i_0\rangle \notin \mathrm{Dom}(p)$. Define the following extension q of p:

$$q = p \cup \{\langle\langle m,i_0\rangle,1\rangle ; m \in \omega \wedge r(m) = 1\} \cup \{\langle\langle m,i_0\rangle,0\rangle ; m \in \omega \wedge r(m)=0\}.$$

Identify subsets a of ω with their corresponding characteristic function

$$\chi_a(m) = \begin{cases} 0 \text{ if } m \notin a \\ 1 \text{ if } m \in a \end{cases}$$

then $q \Vdash^* \underline{r} \subseteq \dot{a}_{i_0}$ and hence $q \Vdash^* \dot{a}_{i_0} \in b_{\underline{r}}$. It follows, that in the model $\mathfrak{N}_0$ defined by any complete sequence $\mathcal{R}_0$ which has p as first and q as second element, it holds that $b_r \cap A \neq \emptyset$. This is in contradiction to the assumption, that p (and hence $\mathcal{R}_0$) forces $b_r \cap A = \emptyset$. Thus, every basic open set (i.e. every absolute interval) of 2^ω contains an element of A and A is in $\mathfrak{N}$ a dense subset of 2^ω, q.e.d.

Continuity-Lemma. Let $\Phi(x_1,\ldots,x_n)$ be an $\mathcal{L}$-formula with no free variables other than $x_1,\ldots,x_n$ and suppose that Φ contains none of the symbols $\dot{a}_i$ (for $i \in \omega$), but Φ may contain constants $\underline{x}$ for $x \in \mathfrak{M}$ or the constant $\dot{A}$. Let $g = \langle g_1,\ldots,g_n\rangle$ be a sequence of different members of A. If $\Phi(g_1,\ldots,g_n)$ holds in $\mathfrak{N}$, then there exists a sequence $\langle b_{r_1},\ldots,b_{r_n}\rangle$ of pairwise disjoint basic open sets of 2^ω, such that $g_\nu \in b_{r_\nu}$ (for $1 \leq \nu \leq n$) and the following holds: if $h = \langle h_1,\ldots,h_n\rangle$ is any sequence of different members of A such that $h_\nu \in b_{r_\nu}$ for $1 \leq \nu \leq n$, then

$$\mathfrak{N} \models \Phi(h_1,\ldots,h_n).$$

<u>Proof</u>. Suppose that for sets $g_1,\ldots,g_n \in A$ the sentence $\Phi(g_1,\ldots,g_n)$ holds in $\mathfrak{N}$. Let $t_1,\ldots,t_n$ be constant terms of $\mathcal{L}$, such that $g_\nu = \mathrm{val}(t_\nu)$ for $1 \leqslant \nu \leqslant n$. Consider the following $\mathcal{L}$-sentence:

$$(0)\quad \left\{ \begin{array}{l} (\Phi(t_1,\ldots,t_n) \wedge t_1 \in \dot{A} \wedge \ldots \wedge t_n \in \dot{A}) \rightarrow \\ \bigvee_{r_1} \cdots \bigvee_{r_n} [\bigwedge_{\nu=1}^{n} (r_\nu \in \bigcup_{k=0}^{\omega} 2^k) \wedge \text{ the basic open intervals} \\ b_{r_\nu} \text{ (for } 1 \leqslant \nu \leqslant n) \text{ are pairwise disjoint} \wedge \bigwedge_{\nu=1}^{n} t_\nu \in b_r \wedge \\ \bigwedge_{x_1} \cdots \bigwedge_{x_n} (\bigwedge_{\nu=1}^{n} (x_\nu \in b_{r_\nu} \wedge x_\nu \in \dot{A}) \rightarrow \Phi(x_1,\ldots,x_n))]. \end{array} \right.$$

The continuity lemma is proved as soon as we have verified that (0) holds in $\mathfrak{N}$. So suppose that the sentence (0) does not hold in $\mathfrak{N}$. Then the negation of (0) holds in $\mathfrak{N}$, and since everything which holds in $\mathfrak{N}$ is forced by some condition of the complete sequence $\mathfrak{R}$ (which defines $\mathfrak{N}$), there exists $p \in \mathfrak{R}$, such that $p \Vdash^* \neg\,(0)$. The statement (0) has the form $\Psi_1 \rightarrow \Psi_2$. Hence $p \Vdash^* \neg\,(0)$ is equivalent to $p \Vdash^* \Psi_1 \wedge \neg\Psi_2$. Thus $p \Vdash^* \Psi_1$ and $p \Vdash^* \neg\,\Psi_2$. We shall obtain a contradiction by showing that there exists an extension p_0 of p such that $p_0 \Vdash^* \Psi_2$. First, $p \Vdash^* \Psi_1$ is:

$$(*)\qquad p \Vdash^* \Phi(t_1,\ldots,t_n) \wedge \bigwedge_{\nu=1}^{n} t_\nu \in \dot{A}$$

It follows from the forcing definition, that there are $i_1,\ldots,i_n \in \omega$ such that $p' \Vdash t_1 = \dot{a}_{i_1} \wedge \ldots \wedge t_n = \dot{a}_{i_n}$ for some extension p' of p. Hence:

$$(**)\quad p' \Vdash^* \Phi(\dot{a}_{i_1},\ldots,\dot{a}_{i_n}).$$

Extend p' to a further condition p'' such that

(1) If $\langle m,i\rangle \in \mathrm{Dom}(p'')$ and $m' < m$, then $\langle m',i\rangle \in \mathrm{Dom}(p'')$,

(2) If $1 \leqslant \nu \leqslant n$, then there is $m \in \omega$ such that $\langle m,i_\nu\rangle \in \mathrm{Dom}(p'')$,

(3) If $j_1 \neq j_2$ then there is $m \in \omega$ such that $\langle m,j_1\rangle \in \mathrm{Dom}(p'')$, $\langle m,j_2\rangle \in \mathrm{Dom}(p'')$ and $p''(\langle m,j_1\rangle) \neq p''(\langle m,j_2\rangle)$,

(4) p'' extends p'.

It is possible to find conditions p'' satisfying (1), (2), (3) and (4). We shall not explicitely describe such a condition, but assume that we have obtained such a p''. Define interval designators (i.e. functions from finite proper initial segments of ω into 2) for $1 \leqslant \nu \leqslant n$, called r_ν, by

$$r_\nu(m) = p''(\langle m,i_\nu\rangle)$$

for $m \in \{m';\ \langle m',i_\nu\rangle \in \mathrm{Dom}(p'')\}$. By condition (1), r_ν is a function defined on some initial segment of ω. By condition (2), r_ν is for no ν with $1 \leqslant \nu \leqslant n$ the empty function and by (3) the basic open sets $b_{r_\nu} = \{f \in 2^\omega;\ f \supseteq r_\nu\}$ are pairwise disjoint (in the sense of the meta-language). Hence this is weakly forced by every condition, thus:

$$\emptyset \Vdash^* \bigwedge_{\nu=1}^{n}(\mathrm{Fnc}(r_\nu) \wedge r_\nu \neq \emptyset) \wedge \bigwedge_\nu \bigwedge_\mu (1 \leq \nu, \mu \leq n \wedge \nu \neq \mu \rightarrow b_{r_\nu} \cap b_{r_\mu} = \emptyset). \quad (\alpha)$$

Since $\mathrm{occ}(\Phi(\dot{a}_{i_1},\ldots,\dot{a}_{i_n})) = \{i_1,\ldots,i_n\}$ by our assumption on Φ, the restriction lemma, applied to (**) with p' replaced by p", yields:

$$(\beta) \qquad p_1 \Vdash^* \Phi(\dot{a}_{i_1},\ldots,\dot{a}_{i_n})$$

where p_1 is the restriction of p" to $\{i_1,\ldots,i_n\}$, id est $p_1 = \{\langle\langle m,i\rangle, e\rangle \in p'';\ i \in \{i_1,\ldots,i_n\}\}$. Now we claim that the following holds:

$$(\gamma) \quad p'' \Vdash^* \bigwedge_{x_1}\ldots\bigwedge_{x_n}[\bigwedge_{\nu=1}^{n}(x_\nu \varepsilon \dot{A} \wedge x_\nu \varepsilon b_{\underline{r}_\nu}) \rightarrow \Phi(x_1,\ldots,x_n)].$$

We have to prove that if $t_1',\ldots,t_n'$ are constant term of $\mathcal{L}$ and q is any extension of p" such that $q \Vdash^* \bigwedge_{\nu=1}^{n}(t_\nu' \varepsilon \dot{A} \wedge t_\nu' \varepsilon b_{\underline{r}_\nu})$, then there exists a further extension q' of q such that $q' \Vdash^* \Phi(t_1',\ldots,t_n')$.

Hence suppose terms $t_1',\ldots,t_n'$ and a condition q are given, where q extends p" and q (weakly) forces the conjunction of the statements $t_\nu' \varepsilon \dot{A} \wedge t_\nu' \varepsilon b_{\underline{r}_\nu}$. Then there exists an extension q' of q and numbers $j_1,\ldots,j_n \in \omega$, such that

$$q \subseteq q' \Vdash^* t_1' = \dot{a}_{j_1} \wedge \ldots \wedge t_n' = \dot{a}_{j_n}.$$

Hence $q' \Vdash \dot{a}_{j_1} \varepsilon b_{r_1} \wedge \ldots \wedge \dot{a}_{j_n} \varepsilon b_{r_n}$. Thus, if we identify subsets of ω with their corresponding characteristic functions :
$q' \Vdash^* \underline{r_1} \subseteq \dot{a}_{j_1} \wedge \ldots \wedge \underline{r_n} \subseteq \dot{a}_{j_n}$. This implies $q_1 \subseteq q'$, where

$$q_1 = \{\langle\langle m,j_\nu\rangle, e\rangle;\ 1 \leq \nu \leq n \wedge r_\nu(m) = e\}.$$

Define the following permutation σ of ω: $\sigma(i_\nu) = j_\nu$ for $1 \leq \nu \leq n$, $\sigma(j_\nu) = i_\nu$, $\sigma(i) = i$ for all other natural numbers i. It follows that

$$q_1 = \sigma(p_1)$$

Hence $p_1 \subseteq p'' \subseteq q \subseteq q'$ and $\sigma(p_1) = q_1 \subseteq q'$, and it follows that $p_1 \cup \sigma(p_1)$ is contained in the condition q'. Thus, $p_1 \cup \sigma(p_1)$ is a condition. Applying σ to (β) we get by the symmetry-lemma

$$\sigma(p_1) = q_1 \Vdash^* \Phi(\dot{a}_{j_1},\ldots,\dot{a}_{j_n})$$

Hence, by the extension lemma $q' \Vdash^* \Phi(\dot{a}_{j_1},\ldots,\dot{a}_{j_n})$. Now, since q' forces $t_1' = \dot{a}_{j_1} \wedge \ldots \wedge t_n' = \dot{a}_{j_n}$ we infer that q' also (weakly) forces $\Phi(t_1',\ldots,t_n')$. This proves (γ).

Both, (α) and (γ), show that p" (weakly) forces that there are non-empty, pairwise disjoint absolute intervals b_{r_ν} such that

$\dot{a}_{i_\nu} \in b_{r_\nu}$, and if $\Phi(\dot{a}_{i_1},\ldots,a_{i_n})$, then $\Phi(t_1'\ldots t_n')$ for every sequence $\langle t_1',\ldots,t_n'\rangle$ with $t_\nu' \in \dot{A} \wedge t_\nu \in b_{r_\nu}$ $(1 \leq \nu \leq n)$. Hence $p'' \Vdash^* \Psi_2$. Since $p \subseteq p' \subseteq p''$, we have obtained the desired contradiction. This proves the continuity-lemma.

Lemma 4. Let r be a function from some finite ordinal into 2. Then $b_r \cap A$ is infinite. Further, A is Dedekind-finite (though infinite in the usual sense).

Proof. By lemma 3 every absolut interval $b_r = \{f \in 2^\omega;\ f \supseteq r\} = \{f \in 2^\omega;\ r_0 \leq f \leq r_1\}$ contains a generic real, where $r_0 = \{\langle m,e\rangle;(m \in Dom(r) \wedge e = r(m)) \vee (m \notin Dom(r) \wedge e = 0)\}$ and $r_1 = \{\langle m,e\rangle;\ (m \in Dom(r) \wedge e = r(m)) \vee (m \notin Dom(r) \wedge e = 1)\}$.
Since every absolut interval includes countably many pairwise disjoint absolute intervals, it follows that $b_r \cap A$ is infinite. Hence A is infinite. That A is Dedekind-finite was proved in section C, page 102. Notice, that the Dedekind-finiteness also follows from the continuity-lemma (see e.g. [35], th.10).

Remark. Since we want to be able to prove that every class V_d of sets u of $\mathfrak{N}$ such that there exists a constant term $E^\alpha x\,\Phi(x)$ with $u = val(E^\alpha x\,\Phi(x))$ and $occ(\Phi) \subseteq d$, where d is a finite subset of ω, has an $\mathfrak{N}$-definable wellordering, we proceed as in the proof of Mathias' theorem (see p.120-121) and define in $\mathfrak{N}$ a ramified language $\mathcal{L}^*$. The alphabet of $\mathcal{L}^*$ contains besides the usual ZF-symbols and the useful limited existential quantifiers V^α and limited comprehension operators E^α (for ordinals α of $\mathfrak{N}$), a name a^* for each $a \in A$, a name A^* for A and names x^* for each $x \in \mathfrak{M}$. Define an interpretation Ω^* for the constant terms of $\mathcal{L}^*$ by induction: $\Omega^*(a^*) = a$, $\Omega^*(A^*) = A$, $\Omega^*(x^*) = x$ and then extending to all limited comprehension terms of $\mathcal{L}^*$. Hence Ω^* is $\mathfrak{N}$-definable (for more details see the analoguous situation on page 121). In order to obtain $\mathfrak{N}$-definable wellorderings of the $\mathfrak{N}$-classes $V_d = \{x \in \mathfrak{N};\ supp^*(x) \subseteq d\}$ we need a support-lemma. The proof of the support-lemma will depend on the following generalization of the continuity-lemma.

Lemma 5. Let $\Phi(x_1,\ldots,x_n)$ be any $\mathcal{L}$-formula with no free variables other than $x_1,\ldots,x_n$ and let $c^* = \{a_i;\ \dot{a}_i$ occurs in $\Phi\}$. If $g_1,\ldots,g_n$ is a sequence of different members of $A-c^*$ and if $\Phi(g_1,\ldots,g_n)$ holds in $\mathfrak{N}$, then there exists a sequence $b_{r_1},\ldots,b_{r_n}$ of absolute intervals of A, pairwise disjoint and disjoint from c^*, such that $g_\nu \in b_{r_\nu}$ $(1 \leqslant \nu \leqslant n)$ and $\Phi(g_1',\ldots,g_n')$ holds for every sequence $g_1',\ldots,g_n'$ of different members of A such that $g_\nu' \in b_{r_\nu}$ for $1 \leqslant \nu \leqslant n$.

Proof. Let $\Psi(x_1,\ldots,x_n, y_1,\ldots,y_m)$ be the formula obtained from Φ by replacing each occurrence of $\dot{a}_{i_\nu}$ by the variable y_ν, where different variables are used for different constants (it is assumed, that the variables $y_1,\ldots,y_m$ do not occur in Φ). If we suppose that $\Phi(g_1,\ldots,g_n)$ holds in $\mathfrak{N}$ for different members of $A-c^*$, then $\Psi(g_1,\ldots,g_n, a_{i_1},\ldots,a_{i_m})$ holds in $\mathfrak{N}$ for different members of A. By the continuity lemma there are pairwise disjoint absolute intervals $b_{r_1},\ldots,b_{r_n}$, $b_{r_{n+1}},\ldots,b_{r_{n+m}}$ of A, such that $\Psi(g_1',\ldots,g_{n+m}')$ holds, whenever $g_\nu' \in b_{r_\nu}$ for $1 \leqslant \nu \leqslant n+m$. Hence $\Phi(g_1',\ldots,g_n')$ holds whenever $g_\nu' \in b_{r_\nu}$ for $1 \leqslant \nu \leqslant n$, since $\Psi(g_1',\ldots,g_n',a_{i_1},\ldots,a_{i_m})$ holds in $\mathfrak{N}$. Since $a_{i_\nu} \in b_{r_{n+\nu}}$ for $1 \leqslant \nu \leqslant m$ and all b_{r_ν} $(1 \leqslant \nu \leqslant n+m)$ are pairwise disjoint, it follows that the b_{r_ν} for $1 \leqslant \nu \leqslant n$ are pairwise disjoint and disjoint from c^*, q.e.d.

Support-lemma. Let $t_1 = E^\alpha x\, \Phi_1(x)$ and $t_2 = E^\alpha x\, \Phi_2(x)$ be limited comprehension terms, and suppose that $\mathfrak{N} \models t_1 = t_2$. There exists an $\mathcal{L}$-formula $\Phi_3(x)$ such that $occ(\Phi_3) = occ(\Phi_1) \cap occ(\Phi_2)$ and and for $t_3 = E^\alpha x\, \Phi_3(x)$ the following holds: $\mathfrak{N} \models t_1 = t_3$.

Proof. Since $p \Vdash^* t_1 = t_2$ for $p \in \mathcal{R}$, we may assume by the restriction lemma, that $\langle m,i\rangle \in Dom(p)$ only if $i \in occ(\Phi_1) \cup occ(\Phi_2)$. Define

$$c = occ(\Phi_1) \cap occ(\Phi_2),\ d_1 = occ(\Phi_1) - c,$$

and assume (without loss of generality), that $d_2 = occ(\Phi_2) - c = \{i_0\}$ has only one element (the general case follows by induction). Split p into subconditions $p_0 = p_0(c) = p/c$, $p_1 = p_1(d_1) = p/d_1$ and $p_2 = p_2(i_0) = p/d_2$. Let $j_0 \in \omega$, $j_0 \notin c \cup d_1 \cup \{i_0\}$ and let τ be a permutation of which leaves $c \cup d_1$ pointwise fixed and maps i_0 onto j_0. Then the symmetry-lemma applied to $p_0(c) \cup p_1(d_1) \cup p_2(i_0) = p$: $p \Vdash^* t_1(c,d_1) = t_2(c,i_0)$ yields

$$p_0(c) \cup p_1(d_1) \cup p_2(j_0) \Vdash^* t_1(c,d_1) = t_2(c,j_0)$$

where $p_2(j_0) = \tau(p_2(i_0))$ and $t_1 = t_1(c,d_1)$, $t_2(c,j_0) = E^\alpha x\, \tau(\Phi(x))$. Hence both relations together, using the restriction lemma entail:

$$p' = p_0(c) \cup p_2(i_0) \cup p_2(j_0) \Vdash^* t_2(c,i_0) = t_2(c,j_0).$$

Obtain an extension p" of p' as in the proof of the continuity lemma, such that p" satisfies conditions (1), (2), (3) and (4) (listed on page 134), where condition (2) reads in the present context as follows:

(2) If $i \in c \cup \{i_0, j_0\}$, then there is $m \in \omega$ such that $\langle m,i\rangle \in Dom(p'')$. Define r_i by: $r_i(m)) : p''(\langle m,i\rangle)$ and $b_{r_i} = \{f \in 2^{\omega};\ r_i \subseteq f\}$. Then the b_{r_i} are pairwise disjoint and $a_i \in b_{r_i}$ for $i \in c \cup \{i_0, j_0\}$. In particular $b_{r_{i_0}} \cap b_{r_{j_0}} = \emptyset$ and $b_{r_{i_0}}$ as well as $b_{r_{j_0}}$ are disjoint from $\{a_i;\ i \in c\}$, see lemma 5.

It was shown in the proof of the continuity-lemma, that

(#) $p'' \Vdash^* \bigwedge_y [y \varepsilon \dot{A} \wedge y \varepsilon b_{r_{\dot{j}_0}} \rightarrow t_2(c,i_0) = t_2(c,y)]$

where $t_1(c,i_0) = t_2(c,x)$ is to be taken as $\Phi(x)$. Now define $\Phi_2^*(y,x)$ to be the $\mathcal{L}$-formula obtained from $\Phi_2(x)$ by replacing $\dot{a}_{i_0}$ by the variable y at all places of occurrence (it is assumed, that y does not occur in $\Phi_2(x)$). Define

$$t_3 = E^{\alpha} x \ (\bigvee_y [y \varepsilon \dot{A} \wedge y \varepsilon b_{r_{i_0}} \wedge \Phi_2^*(x,y)]).$$

By (#): $p'' \Vdash^* t_2(c,i_0) = t_3$, where $occ(t_3) = c$. Since p" can be chosen to be included in $\bigcup \mathcal{R}$, hence in some $q \in \mathcal{R}$, it follows that $q \Vdash t_1 = t_3$, and the lemma is proved.

<u>Lemma 6</u>. The ordering principle (OP) holds in $\mathfrak{N}$.

<u>Proof</u>. For $u \in \mathfrak{N}$ define $supp^*(u)$ as the finite subset of A of minimal cardinality such that there is a term t^* of $\mathcal{L}^*$ mentioning only names a^* for $a \in d$ with $\Omega^*(t^*) = u$. By the support-lemma, $supp^*(u)$ is always defined. Put

$$V_d^0 = \{u \in \mathfrak{N};\ supp^*(u) = d\}$$

then the $\mathfrak{N}$-class V_d has an $\mathfrak{N}$-definable wellordering. Together with the lexicographic ordering of the set of finite subsets of A one concludes, that in $\mathfrak{N}$ every set has a total ordering (for all details see the corresponding proof in section E, page 120-121).

<u>Notice</u>, that every $\mathfrak{N}$-class $V_d = \{u \in \mathfrak{N};\ supp^*(u) \subseteq d\}$ has also an $\mathfrak{N}$-definable wellordering, namely the one inducec (via Ω^*) by the $\mathfrak{N}$-definable well-ordering of the constant terms t of $\mathcal{L}^*$ with $occ(t) \subseteq d$. Notice further, that the proof of lemma 6 shows that more that (OP) holds in $\mathfrak{N}$, namely the Kinna-Wagner ordering principle (KW-O) holds in $\mathfrak{N}$. Next we want to show that also the Boolean Prime Ideal theorem (BPI) holds in $\mathfrak{N}$. First we need the following:

<u>Lemma 7</u>. Let $\Phi(x_0, x_1, \ldots, x_n)$ be a formula of $\mathcal{L}$ with no free variables other than $x_0, \ldots, x_n$ and suppose that $occ(\Phi) \subseteq d$. If $t_1, \ldots, t_n$ are constant terms of $\mathcal{L}$ such that $\bigvee^1_{x_0} \Phi(x_0, t_1, \ldots, t_n)$ holds

in $\mathfrak{N}$ and if $\mathrm{val}(t_\nu) \in V_d$ for $1 \leq \nu \leq n$ and some finite subset d of A, then there is a constant term t_0 of $\mathcal{L}$ such that $\Phi(t_0, t_1, \ldots, t_n)$ holds in $\mathfrak{N}$ and $\mathrm{val}(t_0) \in V_d$.

Proof. Since $\bigvee_{x_0} \Phi(x_0, t_1, \ldots, t_n)$ holds in $\mathfrak{N}$, there is a constant term t' of $\mathcal{L}$ such that $\Phi(t', t_1, \ldots, t_n)$ holds in $\mathfrak{N}$ (since sets of $\mathfrak{N}$ are valuations of terms of $\mathcal{L}$). Suppose $\mathrm{val}(t') \notin V_d$. Let $c = \{i_1, \ldots, i_m\}$ be the set of numbers such that $\dot{a}_{i_\nu}$ occurs in t' and $1 \leq \nu \leq m$ and $a_{i_\nu} \notin d$. Hence $c \neq \emptyset$. Let $t' = E^\alpha x\, \Psi(x)$ and let $\Psi^*(x, y_1, \ldots, y_m)$ be the formula obtained from $\Psi(x)$ by replacing the constants $\dot{a}_{i_\nu}$ by y_ν for $1 \leq \nu \leq m$, different variables for different constants. Hence $t' = E^\alpha x\, \Psi^*(x, \dot{a}_{i_1}, \ldots, \dot{a}_{i_m})$. By lemma 5 we find absolut intervals $b_{r_1}, \ldots, b_{r_m}$ such that $a_{i_\nu} \in b_{r_\nu}$ $(1 \leq \nu \leq m)$ and by lemma 3 generic sets a_{j_ν} such that $a_{j_\nu} \in b_{r_\nu}$ and $a_{j_\nu} \neq a_{i_\nu}$ $(1 \leq \nu \leq m)$ and further the b_{r_ν} are pairwise disjoint and all are disjoint from d. Hence $\mathfrak{N} \models E^\alpha x\, \Psi^*(x, a_{i_1}, \ldots, a_{i_m}) = E^\alpha x\, \Psi^*(x, a_{j_1}, \ldots, a_{j_m})$ and by the support lemma there is a term t_0 such that $\mathfrak{N} \models t' = t_0$ and t_0 mentions only generic reals in d, hence $\mathrm{val}(t_0) \in V_d$, and lemma 7 is proved.

Lemma 8. If $\mathbb{B}$ is a Boolean algebra in $\mathfrak{N}$, then there is a prime ideal J of $\mathbb{B}$ in $\mathfrak{N}$ such that

$$\mathrm{supp}^*(J) \subseteq \mathrm{supp}^*(B).$$

Hence the (BPI) holds in $\mathfrak{N}$.

Proof. Let $\mathbb{B} = \langle B, \sqcap, - \rangle$ be a boolean algebra in $\mathfrak{N}$, where $\sqcap$ is the meetoperation (i.e. product, or greatest lower bound) and $-$ is the complementation operation ($\sqcup$ is definable by means of $\sqcap$ and $-$). Let $d = \mathrm{supp}^*(\mathbb{B})$, hence $\mathbb{B} \in V_d$. Since the operations: projection to the 1st (2nd, 3rd resp.) coordinate, are single valued it follows from lemma 7, that $B, \sqcap$ and $-$ are sets of V_d. Further 1_B and 0_B (the largest and the smallest, resp.) are in V_d, since they are unique.

Consider the set Z of all proper ideals of $\mathbb{B}$, which are in V_d. Since $\{0_B\} \in Z$, $Z \neq \emptyset$. Since V_d has an $\mathfrak{N}$-definable wellordering and $Z \subseteq V_d$, it follows that Z has maximal elements. Let I be a maximal proper ideal of $\mathbb{B}$ with $I \in V_d$ (id est: $I \in Z$). We want to prove that I is prime

Suppose I is not prime. Then for some $b \in B$, $b \notin I$ and $(-b) \notin I$. Since $b \in \mathfrak{N}$, there is a constant term t_b of $\mathcal{L}$, such that $b = \mathrm{val}(t_b)$. We shall derive a contradiction by showing that $1_B \in I$.

It holds that $\mathrm{supp}^*(b) - d \neq \emptyset$, since otherwise the ideal generated by I and $\{b\}$ would be in Z, contradicting the maximality of I.

Hence let t_b a constant term of $\mathcal{L}$ such that $b = val(t_b)$ and, if $t_b = E^{\alpha} x \Psi(x)$ then $occ(\Psi) = \{i \in \omega;\ \dot{a}_i \text{ occurs in } \Psi\}$ is of minimal cardinality. Write

$$occ(\Psi) - d = \{i \in occ(\Psi);\ i \notin d\} = \{i_1, \dots, i_k\}$$

The case $k = 1$ is especially simple, and, as an illuminating example for the proof-procedure, is discussed in detail in Halpern-Lévy [35] and Halpern [31]. We, however, shall start directly with handling the general case, but recommend our reader to look at the discussion of the illuminating example $k = 1$ in [35] and [31].

We shall need a combinatorial theorem of Halpern-Läuchli ([33], theorem 2). Before we formulate a particular case of that theorem, we have to introduce some notation.

A tree $T = \langle T, \leq \rangle$ is a partially ordered set such that for each $x \in T$, $\{y \in T;\ y < x\}$ is totally ordered by $\leq$. The cardinality of this set is called the order of x, or the level at which x occurs. A fan (Fächer) is a non-empty tree such that all elements of it have finite order and each level is a finite set. Hence, if ord(x) is the order of x in the tree $\langle T, \leq \rangle$, then

$\langle T, \leq \rangle$ is a fan $\Leftrightarrow \bigwedge_{x \in T} (ord(x) \in \omega \wedge \bigwedge_{n \in \omega} \{x \in T;\ ord(x) = n\}$ is finite).

Define T/n (restriction of T to n):

$$T/n = \{x \in T;\ ord(x) \leq n\}.$$

Definition. Let D_1 and D_2 be subsets of the fan $\langle T, \leq \rangle$.

D_1 dominates $D_2 \Leftrightarrow \bigwedge_{x \in D_2} \bigvee_{y \in D_1} (x \leq y)$.

D_1 supports $D_2 \Leftrightarrow \bigwedge_{x \in D_2} \bigvee_{y \in D_1} (y \leq x)$.

A subset D of the fan $\langle T, \leq \rangle$ is called (m,1)-dense if there is an element x of T, such that $ord(x) = m$ and $\{y \in T;\ ord(y) = m + 1 \wedge x < y\}$ is dominated by D. Let $\langle T_i, \leq_i \rangle$ be fans for $i < k \in \omega$ and let D_i be a (m,1)-dense subset of T_i (with respect to $\leq_i$) for $i < k$. Then the cartesian product $\prod_{i<k} D_i$ is called an (m,1)-matrix.

Theorem (Halpern-Läuchli). Let $\langle T_i, \leq_i \rangle$, for $i < k$, be finitely many fans without maximal elements. There is a positive integer n such that for every 2-partition $Q = \langle Q_1, Q_2 \rangle$ of $\prod_{i<k} (T_i/n)$, id est $\prod_{i<k} (T_i/n) = Q_1 \cup Q_2$, $Q_1 \cap Q_2 = \emptyset$, either Q_1 or Q_2 includes an (m,1)-matrix for some $m < n$.

For a proof see [33]. (Correct in [33], p.364, the following two misprints: in lemma 1 the second quantifier on the left side of $\models_d$ is an existential quantifier, and two (lines) below in 1.1. the third quantifier on the left side of $\models_d$ is a universal quantifier).

Notice that in [33] and [35] the terminology "finitistic tree" is used. We do not like this philosophically sounding word "finitistic" and use the word "fan" which is also used in Intuitionism.

Now we return to the proof of lemma 8. Let t_B, $t_\sqcap$ and t_- be terms of $\mathcal{L}$ such that $val(t_B) = B$, $val(t_\sqcap) = \sqcap$ (the meet operation of the boolean algebra B) and $val(t_-) = -$ (the complement operation of B), such that $occ(t_B) \subseteq d$, $occ(t_\sqcap) \subseteq d$ and $occ(t_-) \subseteq d$. Further, let $t_b = E^\alpha\, x\, \Psi(x)$ be the term obtained above, and let t_I be the $\mathcal{L}$-term with $occ(t_I) \subseteq d$ such that $val(t_I) = I$. Define:

$$\&(x_1, \ldots, x_k)$$

to be the $\mathcal{L}$-formula obtained from $(t_b \; \varepsilon \; t_B \wedge \neg\, t_b \; \varepsilon \; t_I) \wedge (t_-(t_b) \; \varepsilon \; t_B \wedge \neg\, t_-(t_b) \; \varepsilon \; t_I)$ by replacing each occurrence of $\dot{a}_{i_\nu}$ in t_b (i.e. $E^\alpha\, x\, \Psi(x)$) by the variable x_ν for $1 \leq \nu \leq k$. In the formula used above $t_-(t_b)$ denotes the "complement" of t_b, id est $val(t_-(t_b)) = (-b)$.

Since $\&(\dot{a}_{i_1}, \ldots, \dot{a}_{i_k})$ holds in $\mathfrak{N}$, there exists k absolute intervals $b_{r_1}, \ldots, b_{r_k}$ which are pairwise disjoint and disjoint from d, such that

(1) $$\bigwedge_{\nu=1}^{k} (x_\nu \; \varepsilon \; \dot{A} \wedge x_\nu \; \varepsilon \; b_{r_\nu} \rightarrow \&(x_1, \ldots, x_k))$$

holds in $\mathfrak{N}$ (this follows by lemma 5). We put $S_\nu^0 = \{b_{r_\nu}\}$, for $1 \leq \nu \leq k$, and $S^0 = \langle S_1^0, \ldots, S_k^0 \rangle$. We continue and define k-termed sequences of absolute intervals, S^n, for every $n \in \omega$. Simultanously we prove that the sequences S^n have the following properties (S_ν^n denotes the ν-the coordinate of the sequence S^n):

(P1) S_ν^m is a finite set of absolute subintervals of the members of S_ν^{m-1} for $m \geq 1$ and $1 \leq \nu \leq k$. Therefore, by the definition of S^0 and by induction on m, the members of S_ν^m are subintervals of b_{r_ν} for $1 \leq \nu \leq k$.

(P2) Every member of S_ν^{m-1} has at least two subintervals in S_ν^m for $m \geq 1$ and $1 \leq \nu \leq k$.

(P3) The members of S_ν^m are pairwise disjoint, for $m \geq 0$, $1 \leq \nu \leq k$.

(P4) If $\Sigma \in \Pi\, S^{m-1}$ and G is a finite set of elements of A which contains exactly one member out of each member of $\bigcup \mathrm{Range}(S^m)$, then $\bigsqcap\{val(t_b(u_1, .., u_k)); \langle u_1, .., u_k \rangle \; \varepsilon \prod_{\nu=1}^{k} (\Sigma_\nu \cap G)\} \in I$, and $\bigsqcap\{-val(t_b(u_1, .., u_k)); \langle u_1, .., u_k \rangle \; \varepsilon \prod_{\nu=1}^{k} (\Sigma_\nu \cap G)\} \in I$, where

$t_b(u_1, .., u_k)$ results from t_b by replacing $\dot{a}_{i_\nu}$ by u_ν for $1 \leq \nu \leq k$.

The only one of (P1), ..., (P4) which applies in the case $m = 0$, is (P3); but this holds since the b_{r_ν}'s are pairwise disjoint.

Induction. Let us assume now that for $m \leq n$, S^m is defined and (p1),...,(P4) hold. We shall define S^{n+1} and prove that (P1),... ...,(P4) hold for $m = n+1$. By (P1) the members of S^n_ν are subintervals of b_{r_ν} (for $1 \leq \nu \leq k$). Therefore, if $\Sigma \in \Pi\, S^n$, then $\Pi\Sigma \subseteq \Pi S^*$, where $S^* = \langle b_{r_1},\ldots,b_{r_k}\rangle$. (The symbol Π is used to denote the cartesian product and sequences are understood to be 1-1-functions with domain some element of ω). Hence, if $\Sigma \in \Pi S^n$ then by (1):

$$(2) \quad \Sigma \in \Pi S^n \wedge \langle u_1,\ldots,u_k\rangle \in \Pi(\Sigma_\nu \cap A) \rightarrow \quad \&(u_1,\ldots,u_k)$$

Consider the ideal J generated by $I \cup \{val(t_b(u_1,\ldots,u_k));$ $\langle u_1,\ldots,u_k\rangle \in \Pi(\Sigma_\nu \cap A)\}$. Since $J \in V_d$ and I is a maximal proper ideal in V_d and $I \subseteq J$ we must have $1_B \in J$. Hence, by lemma 1, there is a finite subset $G_1(\Sigma)$ of $\Pi(\Sigma_\nu \cap A)$ such that $\prod\{-val(t_b(u_1,..,u_k));$ $\langle u_1,\ldots,u_k\rangle \in G_1(\Sigma)\} \in I$. By (2) $G_1(\Sigma)$ has at least two members. By considering, in the same way, the ideal generated by $I \cup \{-val(t_b(u_1,\ldots,u_k));\langle u_1,\ldots,u_k\rangle \in \Pi(\Sigma_\nu \cap A)\}$ one obtains a finite subset $G_2(\Sigma)$ of $\Pi(\Sigma_\nu \cap A)$ which has at leat two members such that $\prod\{val(t_b(u_1,\ldots,u_k));\langle u_1,\ldots,u_k\rangle \in G_2(\Sigma)\} \in I$.

Define $G'_\mu(\Sigma)$ to be the set of those elements which occur on some place in an k-tuple of $G_\mu(\Sigma)$, $\mu = 1,2$. $G'_\mu(\Sigma)$ is a finite subset of A and it has at least two members in common with each Σ_ν for $1 \leq \nu \leq k$. Put

$$G^* = \bigcup\{G'_1(\Sigma) \cup G'_2(\Sigma);\ \Sigma \in \Pi S^n\}.$$

Then G^* is a finite subset of A which has at least two members in common with each member of $\bigcup Range(S^n)$. Since $G_\mu(\Sigma) \subseteq \prod_{\nu=1}^{k}(\Sigma_\nu \cap G^*)$ for $1 \leq \mu \leq 2$, we get

$$(3) \quad \Sigma \in \Pi S^n \rightarrow \begin{cases} \prod\{-val(t_b(u_1,..,u_k));\langle u_1,..,u_k\rangle \in \prod_{\nu=1}^{k}(\Sigma_\nu \cap G^*)\} \in I \\ \text{and } \prod\{val(t_b(u_1,..,u_k));\langle u_1,..,u_k\rangle \in \prod_{\nu=1}^{k}(\Sigma_\nu \cap G^*)\} \in I, \end{cases}$$

since $Y_1 \subseteq Y_2 \subseteq B \rightarrow \prod Y_2 \leq \prod Y_1$, and I is an ideal. By lemma 5 we obtain from (3) that there are absolut intervals $b_1,\ldots,b_\lambda$, pairwise disjoint (and disjoint from d) such that, if $G^* = \{a_{j_1},\ldots,a_{j_\lambda}\}$, then $a_{j_\nu} \in b_\nu$ for $1 \leq \nu \leq \lambda$ and

$$(4) \quad \Sigma \in \Pi S^n \wedge \bigwedge_{\nu=1}^{\lambda}[(x_\nu \in \dot{A} \wedge x_\nu \in b_\nu) \rightarrow \prod\{-val(t_b(u_1,\ldots,u_k));\langle u_1,\ldots,u_k\rangle \in \prod_{\nu=1}^{k}(\Sigma_\nu \cap \{x_1,\ldots,x_\lambda\})\} \in I$$
$$\wedge \prod\{val(t_b(u_1,\ldots,u_k));\langle u_1,\ldots,u_k\rangle \in \prod_{\nu=1}^{k}(\Sigma_\nu \cap \{x_1,\ldots,x_\lambda\})\}] \in I.$$

We can assume that each b_ν (for $1 \leq \nu \leq \lambda$) is a subinterval of some member of $\bigcup Range(S^n)$ since this can be attained always by taking appropriate intersections. Moreover it follows from our construction of G^*, that if $s \in \bigcup \mathrm{Range}(S^n)$, then s includes at least two subintervals out of the sequence $\langle b_1, \ldots, b_\lambda \rangle$.

Let us define now the k-sequence S^{n+1} by (for $1 \leq \nu \leq k$):

$$S_\nu^{n+1} = \{b_\gamma;\ 1 \leq \gamma \leq \lambda \wedge b_\gamma \text{ is a subset of a member of } S_\nu^n\}.$$

What we just said concerning the sequence $\langle b_1, \ldots, b_\lambda \rangle$ shows that S^{n+1} satisfied the requirements (P1), P2), (P3) and by (4) also (P4).

Notice, that we defined the infinite sequence $S^0, S^1, \ldots, S^n, \ldots$ by induction in $\mathfrak{N}$ where at each step we made arbitrary choices, namely by selecting $G_\mu(\Sigma)$ for $\mu = 1,2$. But at each step we made only finitely many of them and each set S^n is in V_d which has a definable wellordering (in terms of members of d) as we have shown previously. Hence the construction of $S^0, S^1, \ldots, S^n \ldots$ as given above can be performed inside of $\mathfrak{N}$.

To apply the Halpern-Läuchli theorem we define the following trees.

$$T_\nu = \bigcup_{n\in\omega} S_\nu^n$$

$\leq_\nu$ is the converse of the inclusion relation $\subseteq$.

(for $1 \leq \nu \leq k$). It follows from (P1),...(P4) that the n-th level of $\langle T_\nu, \leq_\nu \rangle$ is exactly S_ν^n and $\langle T_\nu, \leq_\nu \rangle$ is a fan and by (P2) has no tree-tops. Hence all the requirements of the Halpern-Läuchli theorem hold in the present case. Let n be a natural number as in the consequence of that theorem.

Let H be a choice function on the finite set $W = \bigcup\{S_\nu^m;\ m \leq n \wedge 1 \leq \nu \leq k\}$ such that $\mathrm{Range}(H) \subseteq A$. Let y be the k-sequence given by $y_\nu = \{H(s);\ s \in \bigcup\{S_\nu^m;\ m \leq n\}\}$ for $1 \leq \nu \leq k$. We shall show that for every $z \subseteq \Pi y$ either

$$(5)\quad \prod\{\mathrm{val}(t_b(u_1, \ldots, u_k));\langle u_1, \ldots, u_k \rangle \in z\} \in I,$$
$$\text{or } \prod\{-\mathrm{val}(t_b(u_1, \ldots, u_k));\langle u_1, \ldots, u_k \rangle \in z\} \in I.$$

Once (5) is proved then we shall obtain the desired contradiction $1_B \in I$. Let us prove(5)

We define the following 2-partition of $\prod_{\nu=1}^{k}(T_\nu/n)$:

$$Q_1 = \{g \in \prod_{\nu=1}^{k}(T_\nu/n);\langle H(g_1), \ldots, H(g_k)\rangle \in z\}$$

$$Q_2 = \prod_{\nu=1}^{k}(T_\nu/n) - Q_1,$$

where g is the sequence $\langle g_1, \ldots, g_k \rangle$. By our choice of n the Halpern-Läuchli theorem asserts that there exists a positive integer $m_0 < n$ such that either Q_1 or Q_2 includes an $(m_0,1)$-matrix M. Suppose $M \subseteq Q_1$. By definition there are $(m_0,1)$-dense subsets A_ν of T_ν such that

$$M = \prod_{\nu=1}^{k} A_\nu.$$

Since all the sets A_ν are $(m_0,1)$-dense, we may choose a k-sequence of intervals $\langle \tau_1,\dots,\tau_k\rangle$ such that $\tau_\nu \in S_\nu^{m_0}$ and for all $s \in S_\nu^{m_0+1}$, if $s \subseteq \tau_\nu$, then $r \subseteq s$ for some $r \in A_\nu$. With this choice of $\langle \tau_1,\dots,\tau_k\rangle$ define a function f on $\bigcup \mathrm{Range}(S^{m_0+1})$ (using lemma 3) by:

$$f(s) = \begin{cases} \text{if for all } \nu \text{ with } 1 \leq \nu \leq k,\ s \nsubseteq \tau_\nu, \text{ then let } f(s) \text{ be an arbitrary member of } s \cap A. \\ \text{if } (\exists \nu)(1 \leq \nu \leq k \wedge s \subseteq \tau_\nu), \text{ then take an arbitrary member } r \text{ of } A_\nu \text{ for which } r \subseteq s \text{ holds and let } f(s) \text{ be } H(r). \end{cases}$$

Thus for every $s \in \bigcup \mathrm{Range}(S^{m_0+1})$, $f(s) \in s$. Define

$$G = \{f(s);\ s \in \bigcup \mathrm{Range}(S^{m_0+1})\}.$$

By requirement (P4) for m_0+1 we get

(6) $\prod\{\mathrm{val}(t_b(u_1,\dots,u_k)); \langle u_1,\dots,u_k\rangle \in \prod_{\nu=1}^{k}(\tau_\nu \cap G)\} \in I$.

We shall now prove that $\prod_{\nu=1}^{k}(\tau_\nu \cap G) \subseteq z$. Once this is proved, then (5) obviously follows, since $x_1 \leq x_2 \in I_1 \to x_1 \in I$ is a property of ideals. Hence let us prove that the cartesian product of $\tau_\nu \cap G$ is included in z, and let $h = \langle h_1,\dots,h_k\rangle \in \prod_{\nu=1}^{k}(\tau_\nu \cap G)$. Since $h_\nu \in \tau_\nu \cap G$, there are $s_\nu \in \bigcup \mathrm{Range}(S^{m_0+1})$ for $1 \leq \nu \leq k$ such that $f(s_\nu) = h_\nu$. Since f is a choice function, $h_\nu \in s_\nu$.
Since also $h_\nu \in \tau_\nu \in S_\nu^{m_0}$, it follows from (P1) and (P3) that $s_\nu \subseteq \tau_\nu$. But in this case $h_\nu = f(s_\nu) = H(r_\nu)$ where $r_\nu \in A_\nu$. Thus $h = \langle H(r\ ),\dots,H(r_k)\rangle$ where $\langle r_1,\dots,r_k\rangle \in \prod_{\nu=1}^{k} A_\nu \subseteq Q_1$. But by definition of Q_1 this implies $h = \langle H(r_1),\dots,H(r_k)\rangle \in z$ and hence $\Pi(\tau_\nu \cap G) \subseteq z$ holds.

To deal with the other case, namely $M \subseteq Q_2$, let us write $z^* = (\Pi y) - z$ and proceed exactly as in the case $M \subseteq Q_1$, replacing z by z^* and Q_1 by Q_2. Where (P4) was used to obtain (6), we use (P4) now to obtain $\prod\{-\mathrm{val}(t_b(u_1,\dots,u_k));\dots\} \in I$ and get $\prod\{-\mathrm{val}(t_b(u_1,\dots,u_k)); \langle u_1,\dots,u_k\rangle \in z^*\} \in I$, so that (5) holds.

Thus we have shown that for every $z \subseteq \Pi y$, (5) holds. This will be used to obtain by means of lemma 2 the desired contradiction that 1_B would be in I.

Let P be the set of all functions ϕ defined on

$$X = \{\mathrm{val}(t_b(u_1,\dots,u_k)); \langle u_1,\dots,u_k\rangle \in \Pi y\}$$

such that for $x \in X$, $\phi(x) \in \{x,-x\}$. Consider

$$z = \{\langle u_1,\dots,u_k\rangle \in \Pi y;\ \phi(\mathrm{val}(t_b(u_1,\dots,u_k)) = \mathrm{val}(t_b(u_1,\dots,u_k))\}.$$

Then $z \subseteq \Pi y$ and by (5): $\prod\{\mathrm{val}(t_b(u_1,\dots,u_k)); \langle u_1,\dots,u_k\rangle \in z\} \in I$ or $\prod\{-\mathrm{val}(t_b(u_1,\dots,u_k)); \langle u_1,\dots,u_k\rangle \in z\} \in I$. Since one of these elements is in I, their intersection is in any case in I. Thus $\prod\{\phi(x);\ x \in X\} \in I$ for every ϕ. Since P is finite, the union of these intersections is again in I. It was shown in lemma 2, that this union equals 1_B. Hence $1_B \in I$. The assumption, that I is not prime leads, hence, to a contradiction, id est: I is prime. This proves lemma 8.

This proves the theorem, that $\mathfrak{M}[a_0, a_1, \ldots, A] \models ZF + (BPI) + \neg(AC)$. The axiom of choice (AC) is therefore not provable from (BPI) in ZF.

We may use the model constructed above in order to obtain a further independence result. We consider the following two definitions of continuity:

<u>Definition</u> (L.Cauchy): The function f from reals to reals is continuous at x_0 iff for every $\varepsilon > 0$ there is a $\delta > 0$ such that $|x - x_0| < \delta$ implies $|f(x) - f(x_0)| < \varepsilon$.

<u>Definition</u> (Heine - Borel): f is continuous at x_0 iff $\lim_{n\to\infty} f(x_n) = f(x_0)$ for every sequence $\{x_n\}_{n\in\omega}$ convergent to x_0.

In elementary analysis one proves, that both definition are equivalent, but the proof uses the axiom of choice. That the equivalence is no longer true if we drop (AC) has been discovered by Halpern-Lévy in [35] and independently by

[37] M.JAEGERMANN: The axiom of choice and two definitions of Continuity. Bull.Acad.Polon.Sci.vol.13(1965)p.699-704.

<u>Theorem</u> (Halpern-Lévy; Jaegermann): It is not provable in ZF, that every function from reals to reals which is continuous (at x_0) in the sense of Heine-Borel is also continuous (at x_0) in the sense of Cauchy.

Notice, that it is obviously provable in ZF that every Cauchy-continuous function is also Heine-Borel continuous.

<u>Proof</u>. By lemma 4 (see page 136) A is dedekind-finite, while infinite and by lemma 3 (see p.133) A is a dense subset of 2^ω (in the sense of the product topology). The function

$$\phi : x \mapsto \sum_{n=0}^{\infty} \frac{1}{2^{n+1}}$$

for $x \in A$, is a one-one mapping of A into the interval $]0,1]$ (right closed, left open). ϕ is one-to-one since no $x \in A$ is finite (finite subsets of ω are all in the groundmodel $\mathfrak{M}$). Since absolut intervals of A are non-empty, the image $\phi(A)$ is a dense subset of $[0.1]$, and

$\phi(A)$ is a dedekind-finite, while infinite, subset of the closed interval $[0,1]$. Define the following function f:

$$f(x) = \begin{cases} 0 \text{ if } x \in \phi(A) \\ 1 \text{ otherwise} \end{cases}$$

for $x \in [0,1]$. Then $f(0) = 1$, since $0 \notin \phi(A)$. Since $\phi(A)$ is dedekind-finite, every sequence $\{x_n\}_{n\in\omega}$ convergent to 0, for $0 < x_n < 1$, can contain at most finitely elements from $\phi(A)$. Hence $\lim(f(x_n)) = 1 = f(0)$ and f is continuous in the sense of Heine-Borel. Since $\phi(A)$ is dense in $[0,1]$, f is obviously not continuous in the sense of Cauchy. q.e.d.

H) THE AXIOM OF DEPENDENT CHOICES

In his paper "Axiomatic and algebraic aspects of two theorems on sums of cardinals" (Fund.Math.35(1948)p.79-104, in particular p.96) A.Tarski considered the following axiom, which was first formulated by P.Bernays (J.S.L. 7 (1942)p.86):

(DC) AXIOM OF DEPENDENT CHOICES: Let R be a binary relation on the set x such that $(\forall y \in x)(\exists z \in x)(\langle y,z\rangle \in R)$, then there exists a countable sequence $y_0, y_1, \ldots, y_n, \ldots$ $(n \in \omega)$ of elements of x such that $\langle y_n, y_{n+1}\rangle \in R$ for all $n \in \omega$.

The name "dependent choices" is used, since (DC) asserts that there exists a choice sequence where y_n is chosen in dependence of the choice of y_{n-1}. Bernays mentions that (DC) follows from (AC) and that (DC) implies the countable axiom of choice (AC^ω) obviously (see p.100 of these notes for the formulation of (AC^ω)).

Both axioms, (AC^ω) and (DC), are powerful weakened forms of (AC); e.g. in analysis (AC^ω) is sufficient to prove most of the "positive" results such as the first fundamental theorem of Lebesgue-measure. In addition, (DC) is sufficient to prove such results as the Baire category theorem. Further, we mention, that A.Lévy has shown in his paper "A Hierarchy of formulas in set theory" [48], that (DC) is equivalent (in ZF) with some forms of the Löwenheim-Skolem theorem (see [48] p.72-74).

In the formulation of (DC) the choice of y_n is made in dependence of the choice of y_{n-1}. In this formulation, (DC) can not be generalized to yield the existence of sequences of length larger than ω (if certain hypothesis are fullfilled), since e.g. y_ω can not depend on a "predecessor". But we get the idea to let depend y_ω on the set

$\{y_n;\ n \in \omega\}$. More generally we formulate for cardinal numbers α (i.e. finite or an aleph):

(DC^α) Dependent Choices: Let R be a binary relation between subsets and elements of a set x, such that for every $y \subsetneq x$ with $card(y) < \alpha$ there is an $z \in x$ with $\langle y,z \rangle \in R$, then there is a function $f : \alpha \to x$ such that $\langle f''\beta, f'\beta \rangle \in R$ for every ordinal $\beta < \alpha$.

Here $f'\beta = f(\beta)$ and $f''\beta = \{f(\gamma);\ \gamma < \beta\}$. The formulation of ($DC^\alpha$) is due to A.Lévy:

[52] A.LÉVY: The Interdependence of certain consequences of the axiom of choice; Fund.Math. 54(1964)p.135-157.

Lemma 1: Lévy's axiom (DC^ω) is in ZF equivalent with Bernays' axiom of dependent choices (DC).

Proof. (1) Suppose Bernays' (DC) and let R be a binary relation defined between subsets and elements of a set x such that $(\forall y \subset x)$ $(card(y) < \omega \to (\exists z \in x)(\langle y,z \rangle \in R)$. Consider $S = \{u;\ u$ is a finite subset of $x\}$ and define the following relation R^* on S:

$$\langle u_1, u_2 \rangle \in R^* \Leftrightarrow (\exists z \in x)(u_1 \cup \{z\} = u_2 \wedge \langle u_1, z \rangle \in R).$$

By Bernays' (DC) there exists a sequence $u_0, u_1, \ldots, u_n, \ldots$ $(n \in \omega)$ of elements of S such that $\langle u_n, u_{n+1} \rangle \in R^*$ for all $n \in \omega$. Define a function $f : \omega \to x$ by: $f(n)$ is the only element of $u_{n+1} - u_n$. then f satisfies Lévy's (DC^ω).

(2) Now suppose (DC^ω) and let R be a relation on x such that for every $y \in x$ there is $z \in x$ such that $\langle y,z \rangle \in R$. Define

$$S_n = \{\langle z_0, \ldots, z_n \rangle;\ \langle z_0, z_1 \rangle \in R \wedge \langle z_1, z_2 \rangle \in R \wedge \ldots \wedge \langle z_{n-1}, z_n \rangle \in R\}$$

and $S = \bigcup\{S_n;\ n \in \omega\}$; Define the following relation R^* between subsets T and elements t of S by:

$$\langle T,t \rangle \in R^* \Leftrightarrow (\exists z \in x)(\exists t_0 \in T)(t = t_0 * \langle z \rangle)$$

where $t_0 * \langle z \rangle$ denotes concatenation, i.e. $t_0 * \langle z \rangle = \langle v_0, \ldots, v_n, z \rangle$ if $t_0 = \langle v_0, \ldots, v_n \rangle$. By ($DC^\omega$) there is a function $f : \omega \to S$ such that for all $n \in \omega$: $\langle f''n, f'n \rangle \in R^*$. Since $f(0) = f'0 = \langle v_0, \ldots, v_k \rangle \in S$ define $g : \omega \to x$ by $g(0) = v_0, \ldots, g(k) = v_k, g(n) =$ the last coordinate of $f(n-k)$, for $n \geq k+1$. Then g satisfies Bernays' (DC), q.e.d.

Lemma 2 (A.Lévy [52]): $ZF^0 \vdash \bigwedge_\alpha (DC^\alpha) \to (AC)$.

Proof. Let x be any set. Define $W = \{y \subseteq x;\ y$ is wellorderable$\}$ and a binary relation R between subsets of x and elements of x by:

$\langle y,z\rangle \in R \Leftrightarrow (y \in W \wedge z \in x \wedge z \notin y)$.

Let $\alpha = \aleph(x) = \sup\{\lambda;\ \lambda$ is an ordinal and embeddable into $x\}$. α is a cardinal number. Suppose x is not well-orderable; then R satisfies the hypothesis of (DC^{α}), and hence, by (DC^{α}), there exists a function f from α into x such that $\langle f''\beta, f(\beta)\rangle \in R$ for all ordinals $\beta < \alpha$. By definition of R this implies, that α is embeddable into x. Hence $\alpha \in \aleph(x) = \alpha$, a contradiction, since ordinals are allways well-founded, q.e.d.

Corollary 3. $ZF^{0} \vdash \bigwedge_{\alpha}(DC^{\alpha}) \leftrightarrow (AC)$.

Next we shall discuss another family of weakened forms of the axiom of choice and investigate the interdependences, resp. independences between them and the family of statements $\{(DC^{\alpha});\ \alpha$ a cardinal$\}$. Let α be an ordinal: (AC^{α}): if x is a set of non-empty sets, such that card$(x) = \alpha$, then there exists a function f defined on x such then $f(y) \in y$ for all $y \in x$.

A.Lévy has obtained in [52] (among other things) the following results:

Lemma 4: Let α be an aleph, then (in ZF): $(DC^{\alpha}) \rightarrow (AC^{\alpha})$. Let α_1 and α_2 be alephs such that $\alpha_1 < \alpha_2$, then (in ZF) $(DC^{\alpha_2}) \rightarrow (DC^{\alpha_1})$ and $(AC^{\alpha_2}) \rightarrow (AC^{\alpha_1})$.

For a proof see Lévy [52] p.138, p.140 and p.142.

The following nice result was obtained by R.B.Jensen in 1965 (unpublished). We are grateful to A.R.D.Mathias for telling us Jensen's proof.

Theorem (R.B.Jensen): ZF $\vdash \bigwedge_{\alpha}(AC^{\alpha}) \rightarrow (DC^{\omega})$.

Proof. Suppose, this is not true. Then there exists a set X and a binary relation R on X such that $(\forall y \in X)(\exists z \in X)(\langle y,z\rangle \in R)$, but there is no sequence $x_0, x_1, \ldots, x_n, \ldots$ $(n \in \omega)$, such that $\langle x_n, x_{n+1}\rangle \in R$ for all $n \in \omega$. (Notice, that we consider Bernays' (DC) rather than (DC^{ω})). Consider

$$W = \{Y;\ Y \subseteq X \wedge Y \text{ is wellorderable}\}.$$

Then for all $Y \in W$ we get that $\langle Y, R^{-1} \upharpoonright Y\rangle$ is well-founded and we can introduce a rank-notion ρ_Y on Y by induction on the well-founded relation R^{-1} (restricted to Y). Since $\bigcup W = X$ (since all singletons are in W) we can define for every $z \in X$:

$$\lambda(z) = \sup\{\rho_Y(z);\ Y \in W\}.$$

Step 1: If $z \in X$, then $\omega \leq \lambda(z)$.

Proof. Suppose not; then $\lambda(z) = n < \omega$ for some $z \in X$, and hence $\lambda(z) = \rho_Y(z)$ for some $Y \in W$. Thus there are elements $x_1, \ldots, x_n$ in Y, such that $\langle x_1, x_2\rangle \in R^{-1}$, $\langle x_2, x_3\rangle \in R^{-1}, \ldots, \langle x_{n-1}, x_n\rangle \in R^{-1}$ where $x_n = z$. But for given $x_1 \in X$, there exists $x_0 \in X$ such that $\langle x_1, x_0\rangle \in R$, and hence $\langle x_0, x_1\rangle \in R^{-1}$. Hence for $Y^* = Y \cup \{x_0\}$ we get $\rho_{Y^*}(z) = n+1$, a contradiction.

Step 2: For every $z \in X$ there exists $Y \in W$ such that $\lambda(z) = \rho_Y(z)$.

Proof. For ordinals $\gamma \leq \lambda(z)$ consider the following sets:

$$T_\gamma = \{\langle Y,M\rangle;\ Y \in W \wedge M \text{ wellorders } Y \wedge \rho_Y(z) = \gamma\}$$

and define $K(z) = \{T_\gamma;\ \gamma \leq \lambda(z)\}$. The set K(z) is wellordered of type $\leq \lambda(z) + 1$. Hence by $(AC^{\lambda(z)+1})$ there exists a function f which selects from each T_γ one element. Write $f(T_\gamma) = \langle Y_\gamma, M_\gamma\rangle$, and define $Y^* = \bigcup\{Y_\gamma;\ \gamma \leq \lambda(z)\}$. We claim that Y^* is wellorderable. In fact, put $Y_0 = A_0$, $A_1 = Y_1 - Y_0, \ldots, A_\gamma = Y_\gamma - (\bigcup\{Y_\beta;\ \beta < \gamma\}), \ldots$ for $\gamma \leq \lambda(z)$ and $N_0 = M_0, \ldots, N_\gamma = M_\gamma \cap (A_\gamma \times A_\gamma)$; then N_γ wellorders A_γ, and $Y^* = \bigcup\{A_\gamma;\ \gamma \leq \lambda(z)\}$ and the A_γ are pairwise disjoint. Define

$$M^* = \{\langle a,b\rangle;\ \bigvee_\gamma \bigvee_\delta (a \in A_\gamma \wedge b \in A_\delta \wedge [\gamma < \delta \vee (\gamma = \delta \wedge \langle a,b\rangle \in N_\gamma)])\}.$$

then M^* wellorders Y^*. Hence $Y^* \in W$, and $\rho_{Y^*}(z)$ is defined. It follows that $\rho_{Y^*}(z) = \lambda(z)$, q.e.d.

Step 3. The final step: Take an element $z_0 \in X$ such that $\lambda(z_0)$ is minimal in $\{\lambda(z);\ z \in X\}$. By the Hypothesis on R, there exists $z_1 \in X$ such that $\langle z_0, z_1\rangle \in R$. For any $Y \in W$ such that $z_1 \in Y$ define $Y^+ = Y \cup \{z_0\}$. Then always $\rho_{Y^+}(z_0) = \rho_{Y^+}(z_1) + 1$. By the result of step 2, there exists $Y \in W$, such that $\lambda(z_1) = \rho_Y(z_1)$. Hence for this Y:

$$\lambda(z_1) = \rho_{Y^+}(z_1) < \rho_{Y^+}(z_1) + 1 = \rho_{Y^+}(z_0) \leq \lambda(z_0).$$

A contradiction! $\lambda(z_0)$ would not be minimal! This proves Jensen's theorem.

Independence Results

Mostowski has shown in 1948 by means of models containing urelements that (DC) does not imply (AC) in ZF^0, or even: $ZF^0 \nvdash (DC^\omega) \rightarrow (AC^{\omega_1})$:

[68] A.MOSTOWSKI: On the principle of dependent choices; Fund. Math.35 (1948)p.127-130.

We shall describe briefly Mostowski's model. Consider a set theory in which there are $\aleph_1$ many reflexive sets. Let A be the set, of reflexive sets $x = \{x\}$, of cardinality $\aleph_1$. Define $R_0 = A$, $R_\gamma = \bigcup\{P(R_\beta);\ \beta < \gamma\}$ and $V = \bigcup_\gamma R_\gamma$. Take an enumeration of the set of reflexive sets (called "atoms" in the sequel): $R_0 = A = \{a_\gamma; \gamma < \omega_1\}$. Every ordinal γ can be written (in a unique way) as $\beta + n$, where β is a limit ordinal and $n \in \omega$. Write $\gamma \equiv 0$ iff $\gamma = \beta + n$ with n even, and write $\gamma \equiv 1$ iff $\gamma = \beta + n$ with n odd. A permutation f of R_0 (i.e. one-one-mapping of R_0 onto R_0) is called admissible iff f preserves pairs, id est:

Def. f admissible $\Leftrightarrow \bigwedge_{\gamma<\omega_1} [f(a_\gamma) \neq a_\gamma \rightarrow [(\gamma \equiv 0 \wedge f(a_\gamma) = a_{\gamma+1}) \vee (\gamma \equiv 1 \wedge f(a_\gamma) = a_{\gamma-1})]]$.

Hence admissible permutations on R_0 leave $B = \{\{a_\gamma, a_{\gamma+1}\}; \gamma \in \omega_1 \wedge \gamma \equiv 0\}$ pointwise fixed. Let G be the group of all admissible permutations on R_0. Call a subgroup H of G a countable support subgroup if there is a countable subset e of R_0 such that $H = \{f \in G;$ f leaves e pointwise fixed$\}$. Let F be the filter of subgroups of G which has the set of countable support subgroups as filter basis. Define $\mathfrak{M} = \mathfrak{M}[G,F]$ as in chapter III, section B, page 54. Then $\mathfrak{M}$ obviously violates (AC), since e.g. $B \in \mathfrak{M}$, but B has no choice set in $\mathfrak{M}$. Moreover (AC^{ω_1}) is false in $\mathfrak{M}$, since B is wellorderable in $\mathfrak{M}$ of type ω_1. Mostowski shows that (DC), id est (DC^{ω}), holds in $\mathfrak{M}$. In fact, if R is a binary relation in $\mathfrak{M}$, such that R satisfies the hypothesis of (DC). Choose (outside of $\mathfrak{M}$) a countable sequence $x_0, x_1, \ldots$ such that $\langle x_n, x_{n+1}\rangle \in R$ for all $n \in \omega$. Each x_n has a countable support S_n. Since $\bigcup_{n\in\omega} S_n$ is again countable, the subgroup $H = \{f \in G;$ f leaves $\bigcup S_n$ pointwise fixed$\}$ is in F and fixes the sequence $x_0, x_1, \ldots, x_n, \ldots$ which is hence in $\mathfrak{M}$, q.e.d.

Thus we have proved the following:

Theorem (A.Mostowski [68]): There exists a permutation model of ZF^0 containing atoms in which (DC^{ω}) holds, but (AC^{ω_1}) is violated. Thus (DC) $\rightarrow$ (AC), or better (DC) $\rightarrow$ (AC^{ω_1}), is not a theorem of ZF^0.

A.Lévy's paper [52] contains further independence results. Since Lévy's paper is "pre-Cohen", as Mostowski's [68], the results apply only to ZF^0 and the method is by construction of Fraenkel-Mostowski-

Specker models.

Lévy asks in [52] p.137 and in [49] p.224, whether $(AC^{\omega}) \rightarrow (DC^{\omega})$ is provable. R.B.Jensen has solved this problem in 1965 (unpublished). We are grateful to F.R.Drake for sending us his abstract of Jensen's proof. Jensen presented his result during the Logic-Colloquium 1965 in Leicester. He first gave a permutation model containing urelements in order to illuminate the basic idea in his independence proof, and then translated the method to the construction of a Cohen generic model. We follow Jensen and present first his permutation model. In the sequel we make, of course, the tacit assumption, that ZF^0 is consistent.

<u>Theorem</u> (R.B.Jensen): There exists a permutation model $\mathfrak{M}$ of ZF^0 containing atoms in which (AC^{ω}) holds, while (DC^{ω}) is not true in $\mathfrak{M}$. Hence $(AC^{\omega}) \rightarrow (DC^{\omega})$ is not provable in ZF^0.

<u>Proof</u>. Take a set theory with choice in which there is a set R_0 of reflexive sets (called atoms), such that R_0 has cardinality $\aleph_1$. We want to define a certain tree-ordering on R_0.

Consider first the well-ordered set ω_1 and consider

$$\omega_1^m = \{\langle x_1, \ldots, x_m\rangle ; x_1 < x_2 < \ldots < x_m \in \omega_1\}$$

and $\underset{\sim}{H} = \bigcup\{\omega_1^m ; 1 \leq m < \omega\}$. There is a natural partial ordering $\ll$ on $\underset{\sim}{H}$ defined by: $s_1 \ll s_2 \Leftrightarrow$ the sequence s_1 is an initial segment of s_2. More precisely one defines first $s_1 \lessdot s_2$ to express that s_1 immediately preceeds s_2:

$$s_1 \lessdot s_2 \Leftrightarrow (s_1 = \langle x_1, \ldots, x_m\rangle \wedge s_2 = \langle x_1, \ldots, x_m y\rangle)$$

Hence $s_1 \lessdot s_2 \Leftrightarrow (\exists y \in \omega_1)(s_2 = s_1 * \langle y\rangle)$ if $*$ denotes concatenation of sequence. Now define

$s_1 \ll s_2 \Leftrightarrow$ there are finitely many elements of $\underset{\sim}{H}$, say $h_1, \ldots, h_n$, such that $s_1 = h_1 \lessdot h_2 \lessdot \ldots \lessdot h_n = s_2$.

Since $\underset{\sim}{H}$ has cardinality $\aleph_1$, there exists a one-to-one mapping f from R_0 onto $\underset{\sim}{H}$. Via f one carries the tree-structure $\ll$ over to R_0 by:

$$a < b \Leftrightarrow f(a) \ll f(b).$$

Hence $<$ is a strict partial ordering on R_0 and $\langle R_0, \leq\rangle$ is a tree. Notice that $\langle R_0, \leq\rangle$ has no tree-tops, that the order of each element of R_0 is finite and that for every $a \in R_0$ the set of immediate successors $\{b \in R_0 ; a < b \wedge \mathrm{ord}(b) = \mathrm{ord}(a) + 1\}$ has cardinality $\aleph_1$.

Let G be the group of all orderpreserving one-to-one mappings from R_0 onto R_0. In order to define an interesting filter F of subgroups we define the notion of a "small subtree". First:

B is a <u>complete branch</u> in $\langle R_0, \leq \rangle$ iff $\langle B, \leq \rangle$ is totally ordered and if $a < b \in B$, then $a \in B$ and there does not exists an element $d \in R_0$ such that $a < d$ for all $a \in B$.

<u>Definition</u>. If T is a subset of R_0, then $\langle T, \leq \rangle$ is a small subtree of $\langle R_0, \leq \rangle$ iff T is countable, $\langle T, \leq \rangle$ is a subtree (id est: $a < b \in T$ implies $a \in T$) and no branch in $\langle T, \leq \rangle$ is a complete branch in $\langle R_0, \leq \rangle$.

<u>Definition</u>. A subgroup H of G is called a <u>nice</u> subgroup iff there exists a small subtree $\langle T, \leq \rangle$ such that

$$H = \{\pi \in G;\ \pi \text{ leaves } T \text{ pointwise fixed}\} = K[T]$$

Define F to be the set of those subgroups of G which include a nice subgroup. F is a filter of subgroups. Define $H[x] = \{\pi \in G;\ \pi(x) = x\}$ and let $TC(x) = \{x\} \cup x \cup \bigcup x \cup \ldots$ be the transitive closure of x. Define as in chapter III, p.54: $\mathfrak{M} = \mathfrak{M}[G,F] = \{x;\ \bigwedge_y (y \in TC(x) \rightarrow H[y] \in F)\}$. Specker's theorem (see p.54) shows that $\mathfrak{M}$ is a model of ZF^0. We shall show that $(AC^{\omega}) \wedge \neg (DC^{\omega})$ holds in $\mathfrak{M}$.

<u>Lemma</u>: The axiom of dependent choices (DC^{ω}) does not hold in $\mathfrak{M}$.

<u>Proof</u>. Since for $a \in R_0$, a has finite order, and hence $\{b \in R;\ b \leq a\}$ is a small subtree it follows that $K[a] = H[a] \in F$ and hence R_0, every element of R_0 and the tree-ordering $\leq$ is in $\mathfrak{M}$. Obviously $\leq$ is a binary relation satisfying $(\forall x \in R_0)(\exists y \in R_0)(x < y)$. If (DC^{ω}), and hence (DC), would hold in $\mathfrak{M}$, then there would be a countable sequence $S = \{a_0, a_1, \ldots, a_n, \ldots\}$ of elements of R_0 in $\mathfrak{M}$, such that $a_0 < a_1 < \ldots < a_n < \ldots$. If S is in $\mathfrak{M}$, then there is a small subtree T of R_0 such that if $\pi \in K[T] = \{\sigma \in G;\ \sigma \text{ leaves } T$ pointwise fixed$\}$ (by def.see p.57) entails $\pi(S) = S$, id est $K[T] \leq H[S]$. But $S^* = \{b \in R_0;\ (\exists a \in S)(b \leq a)\}$ is a complete branch in $\langle R_0, \leq \rangle$. Since T is small, $T \cap S^*$ is finite (or even possibly empty). Hence there exists $b_0 \in S$ with $b_0 \notin T$. Define the following permutation τ of R_0: pick any element $c_0 \in R_0 - T$ such that, if b_1 is an immediate predecessor of b_0, then $b_1 < c_0$ and $ord(c_0) = ord(b_1) + 1$ (hence $ord(c_0) = ord(b_0)$ and c_0 and b_0 are in the same set of immediate successors of b_1). Let τ be the identity on $\{x \in R_0;\ \neg(x \geq b_0 \vee x \geq c_0)\}$ and $\tau(b_0) = c_0$, $\tau(c_0) = b_0$ and τ maps the subset $\{x \in R_0;\ x \geq b_0\}$ onto $\{x \in R_0;\ x \geq c_0\}$ and reversely. Hence $\tau \in K[T]$ and since $K[T] \leq H[S]$, it follows that $\tau(S) = S$.

But $b_0 \in S$ implies $c_0 = \tau(b_0) \in \tau(S) = S$, a contradiction. This proves the lemma.

Lemma. The countable axiom of choice (AC^{ω}) holds in $\mathfrak{M}$.

Proof. Let $z = \{x_i ; i \in \omega\}$ be a countable set in $\mathfrak{M}$ such that $i \neq j \to x_i \cap x_j = \emptyset$, $x_i \neq \emptyset$ for all $i \in \omega$ and the sequence $\{\langle x_i, i\rangle ; i \in \omega\} = \Sigma$ is in $\mathfrak{M}$. Hence there is a small subtree T such that $K[T] \leqslant K[\Sigma] \leqq H[\Sigma] \leqslant H[z]$ (since Σ is a wellordering of z, see p.57-58). Write $T = T_z$. Proceed outside of $\mathfrak{M}$. Since (AC) holds in the surrounding set theory, there exists a choice set $C = \{y_i ; i \in \omega\}$ such that $y_i \in x_i$ for all $i \in \omega$. The set C need **not** be in $\mathfrak{M}$, but $C \subseteq \mathfrak{M}$ by the transitivity of $\mathfrak{M}$. We are looking for mappings (not in G) which transform C into some choice set C^* which is in $\mathfrak{M}$.

Since $y_i \in \mathfrak{M}$, there are small subtrees T_i of $\langle R_0, \leqslant\rangle$ such that $K[T_i] \leqslant H[y_i]$ for $i \in \omega$. Proceed by cases.

Case 1. $T^* = \bigcup\{T_i ; i \in \omega\}$ is a small subtree. Then obviously $K[T^*] \leqslant K[C] \leqq K[C]$ and C is in $\mathfrak{M}$ and we are done since we have obtained a choice-set C for z in $\mathfrak{M}$.

Case 2. T^* is not a small subtree. We shall construct a sequence of permutations $\pi_i \in K[T_z]$ such that $T^0 = \bigcup\{\pi_i(T_i) ; i \in \omega\}$ is a small subtree. We construct these permutations π_i by induction on i.

I) Let π_0 be the identity on R_0.
II) Suppose that for $0 \leqslant i < n$ permutations $\pi_i \in K[T_z]$ are defined.
III) Construction of π_n.

Of course, the construction of π_n takes place outside of $\mathfrak{M}$. Remember that the tree structure $\leqslant$ on R_0 has the property, that the set of immediate successors of any element of R_0 has (outside of $\mathfrak{M}$!) cardinality $\aleph_1$ while small subtrees always have cardinality $\leqslant \aleph_0$. Hence we may shift (displace) the tree T_n into a tree $\pi_n(T_n)$ so that $T_z \cap T_n = T_z \cap \pi_n(T_n)$ but $\pi_n(T_n) - T_z$ is disjoint from $\bigcup\{\pi_i(T_i) ; 0 \leqslant i < n\}$. In details:

For $a \in R_0$ let $[a]$ be $\{x \in R_0 ; a \leqslant x\}$. Further call the cardinality of $\{y \in R_0 ; y \leqslant a \wedge y \neq a\}$ the order of a, in symbols ord(a). Hence ord(a) is for $a \in R_0$ always finite. Define $L(k)$ to be the k-th level:

$$L(k) = \{x \in R_0 ; \text{ord}(x) = k\}$$

For $y \in L(k)$ let $I(y)$ be the set of immediate successors of y:

$$I(y) = \{x \in R_0 ; y < x \wedge \text{ord}(x) = \text{ord}(y) + 1\}$$

Thus $y_1 \neq y_2$, $y_1, y_2 \in L(k)$ implies $I(y_1) \cap I(y_2) = \emptyset$ (and $I(y)$) always has cardinality $\aleph_1$ and further $L(k+1) = \bigcup\{I(y);\ y \in L(k)\}$.

When we consider $\langle [a], \leqslant\rangle$ we mean of course the ordered pair consisting of $[a]$ and $\leqslant$ restricted to $[a]$. If $a_1, a_2 \in L(k)$, then $\langle [a_1], \leqslant\rangle$ and $\langle [a_2], \leqslant\rangle$ are isomorphic (and isomorphic with $\langle R_0, \leqslant\rangle$). Let $\sigma^{a_1}_{a_2}$ be such an isomorphism which maps $[a_1]$ one-to-one onto $[a_2]$ in an orderpreserving way. If $\mathrm{ord}(a_1) \neq \mathrm{ord}(a_2)$ then $\sigma^{a_1}_{a_2}$ is undefined.

For $m < \mathrm{ord}(x)$ define $\lambda(m,x)$ to be the element $z \in R_0$ with $\mathrm{ord}(z) = m$ and $z < x$. Since $\leqslant$ is a tree-ordering of height ω, $\lambda(m,x)$ is unique.

With this amount of notation we are able to define by induction (on $k \in \omega$) a sequence of permutations $g^k_n \in G$ which approximate π_n. g^0_n is the identical mapping on R_0. Suppose g^m_n is defined for $0 \leqslant m \leqslant k$. Define g^k_n by means of g^{k-1}_n, σ and the following function ϕ which we are going to characterize:

Since T_z, T_n and $\bigcup\{\pi_i(T_i);\ 0 \leqslant i < n\}$ are all countable and each set $I(y)$ for $y \in L(k-1)$ is uncountable we may find a function ϕ from $L(k)$ onto $L(k)$ which preserves the partition $\{I(y);\ y \in L(k-1)\}$ id est $\phi''I(y) = \{\phi(x);\ x \in I(y)\} = I(y)$ for all $y \in L(k-1)$, and leaves $L(k) \cap T_z$ pointwise fixed and $\{g(x);\ x \in L(k) \cap T_n \wedge x \notin T_z\} \cap \{x \in L(k);\ x \in T_z \cup \bigcup_{i=0}^{n-1} \pi_i(T_i)\} = \emptyset$.

Now define g^k_n : let x be in R_0:

$$g^k_n(x) = \begin{cases} g^{k-1}_n(x) & \text{if } \mathrm{ord}(x) < k, \\ \phi(g^{k-1}_n(x)) & \text{if } \mathrm{ord}(x) = k, \\ \sigma^{\lambda(k,x)}_{\phi(\lambda(k,x))}(x) & \text{if } \mathrm{ord}(x) > k. \end{cases}$$

Thus the sequence g^k_n for $k \in \omega$ is defined by induction for all $k \in \omega$ and we define π_n : let $x \in R_0$:

$$\pi_n(x) = g^{\mathrm{ord}(x)}_n(x)$$

It follows that $\pi_n \in K[T_z]$ and $\pi_n(T_n)$ is "disjoint from $\bigcup_{i=0}^{n-1}\pi_i(T_i)$ modulo T_z". Since we have defined the sequence $\pi_0, \pi_1, \ldots, \pi_n, \ldots$ by induction it follows that every branch B which is included in

$$T^0 = \bigcup\{\pi_i(T_i);\ i \in \omega\}$$

is either included in T_z or in one of the small subtrees $\pi_i(T_i)$. Hence T^0 itself is small. Hence $K[T^0] \in F$.

Define $C^0 = \{\pi_n(y_n);\ n \in \omega\}$; then $K[T^0] \leqslant K[\pi_n(T_n)] \leqslant H[\pi_n(y_n)]$ since $K[T_n] \leqslant H[y_n]$. Thus $K[T^0] \leqslant K[C^0] \leqslant H[C^0]$ and it follows that C^0 is a set of $\mathfrak{M}$.

Since $\pi_n \in K[T_z] \leq K[z]$ it follows from $y_n \in x_n \in z$ that $\pi_n(y_n) \in \pi_n(x_n) = x_n \in z$. Hence C^0 is a choice set. This proves the lemma and Jensen's theorem is established.

Discussion. The function ord can be given in $\mathfrak{M}$ and $\{L(k); k \in \omega\}$ is a countable set in $\mathfrak{M}$. By (AC^ω) in $\mathfrak{M}$, $\mathfrak{M}$ contains a choice set C for this set. It follows that $\{x \in R_0;\ (\exists y \in C)(x \leq y)\}$ is a small subtree!

Another point is, that the method of proof given above can be used to yield the following generalization:

Theorem: If α is an aleph, then $(AC^\alpha) \to (DC^\omega)$ is not provable in ZF^0 (provided ZF^0 is consistent).

The proof is analoguous to the one given above. Instead of defining $\underset{\sim}{H}$ to be $\bigcup\{\omega_1^m;\ m \in \omega\}$ one takes $\bigcup\{(\alpha^+)^m;\ m \in \omega\}$ as $\underset{\sim}{H}$, where α^+ is the successor aleph of α. Small subtrees of $\langle R_0, \leq \rangle$ are subtrees without complete branches of cardinality $\leq \alpha$. Notice, that here R_0 has cardinality α^+. The proof can be carried over to the present case, since α^+ is a regular cardinal.

This generalization shows, that the other result of Jensen, namely $ZF^0 \vdash \bigwedge_\alpha (AC^\alpha) \to (DC^\omega)$ is the "best possible" result.

Translation to a Cohen-generic model

Theorem (R.B.Jensen 1965): Every countable standard model $\mathfrak{M}$ of ZF + V = L can be extended to a countable standard model $\mathfrak{N}$ of ZF such that:

(a) The ordinals of $\mathfrak{N}$ are exactly the ordinals of $\mathfrak{M}$,

(b) the alephs of $\mathfrak{N}$ are exactly the alephs of $\mathfrak{M}$,

(c) (AC^ω) holds in $\mathfrak{N}$,

(d) (DC^ω) does not hold in $\mathfrak{N}$.

Proof. Let $\mathfrak{M}$ be a countable standard model of ZF + V = L. We extend $\mathfrak{M}$ by adding to $\mathfrak{M}$ a generic copy of $\underset{\sim}{H}$ (the tree defined on p.151) and generically all small subtrees T of $\underset{\sim}{H}$ together with wellorderings on each T.

Hence take the first uncountable ordinal $\omega_1^{\mathfrak{M}}$ of $\mathfrak{M}$ and define $\underset{\sim}{H}$ in $\mathfrak{M}$ and the tree-ordering $\leq$ on $\underset{\sim}{H}$ as before. Next define a ramified language $\mathcal{L}$ in $\mathfrak{M}$ which contains besides the usual ZF-symbols, the limited quantifiers $\bigvee^\alpha$, the limited comprehension terms E^α (for ordinals α of $\mathfrak{M}$), constants $\underset{\sim}{x}$ for each set x of $\mathfrak{M}$, constants $\dot{a}_h$ for each $h \in \underset{\sim}{H}$, a binary predicate $\dot{\leqslant}$ and binary predicates $\dot{S}_T$ for each small subtree T of $\underset{\sim}{H}$.

A condition p is a finite partial function from $\omega \times \underset{\sim}{H}$ into 2. Define a strong forcing relation $\Vdash$ containing the following key-clauses:

$p \Vdash t \,\varepsilon\, \dot{a}_h \Leftrightarrow (\exists\, n \in \omega)(p \Vdash t \simeq \underset{\sim}{n} \;\&\; p(\langle n,h\rangle) = 1)$,

$p \Vdash t_1 \,\dot{\leqslant}\, t_2 \Leftrightarrow (\exists\, h_1, h_2 \in \underset{\sim}{H})(h_1 \leqslant h_2 \;\&\; p \Vdash t_1 \simeq \dot{a}_{h_1} \;\&\; p \Vdash t_2 \simeq \dot{a}_{h_2})$,

$p \Vdash \dot{S}_T(t_1, t_2) \Leftrightarrow (\exists\, h \in \underset{\sim}{H})(h \in T \;\&\; p \Vdash t_1 \simeq \dot{a}_h \;\&\; p \Vdash t_2 \simeq \underset{\sim}{h})$.

where t, t_1, t_2 are constant terms of $\mathcal{L}$. Obtain a complete sequence $\mathcal{R}$ of conditions and thereby a valuation val(t) of the constant terms t. By the Hauptsatz, $\mathfrak{N} = \{\mathrm{val}(t);\ t$ constant term of $\mathcal{L}\}$ is a model of ZF which contains $\mathfrak{M}$ as a complete submodel. Further $\mathfrak{M}$ and $\mathfrak{N}$ have precisely the same ordinals (by lemma T, see page 90); hence (a) holds.

Notice that

1) $a_h = \mathrm{val}(\dot{a}_h) \subseteq \omega$,

2) $\leqslant\ = \mathrm{val}\ (\dot{\leqslant}\ )$ is a tree-ordering the field of which is just the set $\{a_h;\ h \in \underset{\sim}{H}\}$,

3) $S_T = \mathrm{val}\ (\dot{S}_T) = \{\langle a_h, h\rangle;\ h \in T \subseteq \underset{\sim}{H}\}$

more precisely $\leqslant$ is the valuation of $E^{\omega+1}x(\bigvee_y \bigvee_z (x = \langle y,z\rangle \wedge y \,\dot{\leqslant}\, z))$ and similar for S_T. Since we took symbols $\dot{S}_T$ into $\mathcal{L}$ only for small subtrees T which are in $\mathfrak{M}$ and $\mathfrak{M}$ satisfies V = L, hence (AC), there is in $\mathfrak{M}$ a wellordering W_T for T. Since S_T is a one-one-function from $\{a_h;\ h \in T\}$ onto T, and val $(\underset{\sim}{W_T}) = W_T$ is in $\mathfrak{N}$ it follows that in $\mathfrak{N}$ the sets $\{a_h;\ h \in T\}$ (T a small subtree of $\underset{\sim}{H}$ in $\mathfrak{M}$) are wellorderable.

Let G be the group (in $\mathfrak{M}$) of all orderpreserving one-to-one-mappings π from $\underset{\sim}{H}$ onto $\underset{\sim}{H}$. For a condition p define

$\pi(p) = \{\langle\langle n,\pi(h)\rangle, e\rangle;\ \langle\langle n,h\rangle, e\rangle \in p\}$

and define the action of G on $\mathcal{L}$ in the following way: if $\pi \in G$ and $h \in \underset{\sim}{H}$, then $\pi(\dot{a}_h) = \dot{a}_{\pi(h)}$, $\pi(\dot{S}_T) = \dot{S}_{\pi''T}$, where

$\pi''T = \{\pi(h);\ h \in T\}$

and for an $\mathcal{L}$-formula Φ let $\pi(\Phi)$ be result obtained from Φ by replacing any occurence of $\dot{a}_h$ by $\pi(\dot{a}_h)$ and of $\dot{S}_T$ by $\pi(\dot{S}_T)$ (the other symbols of Φ remain unchanged). Then the following holds:

<u>Symmetry-Lemma</u>: If $\pi \in G$, p is a condition and Φ an $\mathcal{L}$-sentence, then

$$p \Vdash \Phi \Leftrightarrow \pi(p) \Vdash \pi(\Phi).$$

Proof by induction (see page 99).
This lemma enables us to prove:

Lemma: The countable axiom of dependent choices (DC^{ω}) does not hold in $\mathfrak{N}$.

Proof. One shows that no condition p can force that $\{a_h; h \in \underset{\sim}{H}\}$, partially ordered by $\leq$, has a complete branch. See the proof of $\neg (DC^{\omega})$ in the preceeding FraenkelMostowski-Specker model.

In order to show that (AC^{ω}) holds in $\mathfrak{N}$ we need a lemma which says, that any set of $\mathfrak{M}$ which is $\mathfrak{N}$-countable (i.e. countable in $\mathfrak{N}$) is also $\mathfrak{M}$-countable. More generally we shall prove that cardinals are preserved in the transition from $\mathfrak{M}$ to $\mathfrak{N}$, thus proving (b).

Combinatorial Lemma. Let B be in $\mathfrak{M}$ a set of conditions. There exists in $\mathfrak{M}$ a subset B' of B such that B' is in $\mathfrak{M}$ countable and for every $p \in B$ there is a $p' \in B'$ such that p and p' are compatible.

For a proof see e.g. Jensen's lecture notes [39] (Springer-Berlin) page 147, or these notes page 106-107. The lemma implies obviously that if B is in $\mathfrak{M}$ a set of conditions whose elements are pairwise incompatible, then B is countable. We have shown on p.106-108 (these notes), that this implies that cardinals (i.e. alephs or finite ordinals) are absolut in the extension from $\mathfrak{M}$ to $\mathfrak{N}$; thus we have shown:

Lemma. For every ordinal γ of $\mathfrak{N}$ (and hence of $\mathfrak{M}$) the cardinality of γ in the sense of $\mathfrak{N}$ is equal to the cardinality of γ in the sense of $\mathfrak{M}$:

$$\overline{\overline{\gamma}}^{\mathfrak{M}} = \overline{\overline{\gamma}}^{\mathfrak{N}},$$

hence $\mathfrak{M}$ and $\mathfrak{N}$ have precisely the same alephs.

In order to show that (AC^{ω}) holds in $\mathfrak{N}$ we introduce the following notation: for a small subtree T of $\langle \underset{\sim}{H}, \leq \rangle$ let V(T) be the $\mathfrak{N}$-class of sets which are explicit definable from T. Since $\mathfrak{L}$ is $\mathfrak{M}$-definable and $\mathfrak{M}$ is a complete submodel of $\mathfrak{N}$ (see the necessary remark on p.96-97), $\mathfrak{L}$ is $\mathfrak{N}$-definable. Since the correspondence $h \to a_h$ is not in $\mathfrak{N}$ (for all $h \in \underset{\sim}{H}$), we cannot interpret $\mathfrak{L}$ inside of $\mathfrak{N}$, but what we can do is to interpret certain sublanguages $\mathfrak{L}(T)$ of $\mathfrak{L}$

in $\mathfrak{N}$. Namely let for a small subtree T be $\mathcal{L}(T)$ the language which contains besides the ZF-symbols, the symbols V^α and E^α, further $\underline{x}$ for $x \in \mathfrak{M}$ and $\dot{\leqslant}$, only constants $\dot{a}_h$ for $h \in T$ and symbols $\dot{S}_D$ for small subtrees D for $D \subseteq T$.

Notation: we say that the set s of $\mathfrak{N}$ is explicit definable from T if there is a constant term t_s of $\mathcal{L}(T)$ such that $s = val(t_s)$.

Now let V(T) be the collection of all sets of $\mathfrak{N}$ which are explicit definable from T. We claim that V(T) is $\mathfrak{N}$ -definable and has an $\mathfrak{N}$ -definable wellordering.

In fact, since $\mathcal{L}(T)$ is $\mathfrak{M}$-definable, hence $\mathfrak{N}$ -definable, and the correspondences $a_h \to h$ for $h \in T$ and $S_D \to D$ (since $D = Range(S_D)$) for $D \subseteq T$ are in $\mathfrak{N}$ we can define an interpretation Ω for constant terms t of $\mathcal{L}(T)$ in $\mathfrak{N}$ by setting

$$\Omega(\dot{a}_h) = a_h,\ \Omega(\dot{S}_D) = S_D \quad \text{for } h \in T,\ D \subseteq T,\ D \text{ small subtree,}$$

$$\Omega(\dot{\leqslant}) = \leqslant \text{ and } \Omega(\underline{x}) = x,$$

and then extending by induction to all t's of $\mathcal{L}(T)$. Thus

$$V(T) = \{\Omega(t); t \text{ is a constant term of } \mathcal{L}(T)\}$$

and V(T) is an $\mathfrak{N}$ -definable class.

Lemma. If T is a small subtree of $\langle \underset{\sim}{H}, \leqslant \rangle$, then V(T) has an $\mathfrak{N}$ -definable wellordering.

Proof. Since T and $\{D \subseteq T;\ D$ a small subtree$\}$ are wellorderable in $\mathfrak{M}$ and further $\{\underline{x};\ x$ a set of $\mathfrak{M}\}$, $\{V^\alpha;\ \alpha \in On^{\mathfrak{M}}\}$ and $\{E^\alpha;\ \alpha \in On^{\mathfrak{M}}\}$ have $\mathfrak{M}$-definable wellorderings, it is clear, that the alphabet of $\mathcal{L}(T)$ is wellorderable in $\mathfrak{M}$. Hence the class of all constant terms t of $\mathcal{L}(T)$ (considered as finite sequences of symbols from the alphabet of $\mathcal{L}(T)$) has a (lexicographic) wellordering which is carried over to V(T) via Ω.

Notice, that Ω can not be defined for all terms of $\mathcal{L}$, but only for terms of $\mathcal{L}(T)$; but this is sufficient in the present case, q.e.d.

Lemma. For every term t of $\mathcal{L}$ there is a small subtree T such that for every condition p it holds that $p \Vdash^* t \neq \emptyset \to t \cap C(T) \neq \emptyset$.

Sketch of Proof. C(T) is the unlimited term of $\mathcal{L}$ so that $V(T) = val(C(T))$. For given t consider $B = \{q \in Cond;\ q \Vdash^* \bigvee_x x \in t\}$, where Cond is the $\mathfrak{M}$-set of conditions. Now apply the combinatorial lemma and obtain a $\mathfrak{M}$-countable subset B' of B so that for every $q \in B$ there is $q' \in B'$ with $q \cup q' \in Cond$, and construct the tree T. Then for $p' \in B'$ there is t' of $\mathcal{L}(T)$ so that $p' \Vdash^* t \neq \emptyset \to t' \in t$.

Lemma. (AC^{ω}) holds in $\mathfrak{N}$.

Sketch of proof. Let z be a countable set in $\mathfrak{N}$, $z = \{z_i; i \in \omega\}$ so that each $z_i \in z$ is not empty. Let t_z, t_i $(i \in \omega)$ be constant terms of $\mathcal{L}$ so that $z = val(t_z)$ and $z_i = val(t_i)$. By the preceeding lemma there is a sequence of small subtrees T_i such that $\emptyset \Vdash^* t_i \neq \emptyset \to t_i \cap C(T_i)$. Obtain a sequence of permutations π_i leaving t_z invariant but such that $T^* = \bigcup\{\pi_i(T_i); i \in \omega\}$ is a small subtree (use the construction presented in the Fraenkel-Mostowski-Specker version of the model). Hence in $\mathfrak{N}$: $z_i \cap V(T^*) \neq \emptyset$ for all $i \in \omega$. Since $V(T^*)$ has an $\mathfrak{N}$ -definable wellordering we obtain a choice sequence in $\mathfrak{N}$. This proves Jensen's theorem. Again, as in the case of the permutation model, the theorem can be strengthened to

Theorem (Jensen): Let α be an infinite cardinal in a countable standard model $\mathfrak{M}$ of ZF + V = L; then $\mathfrak{M}$ can be extended to a countable standard model $\mathfrak{N}$ of ZF in which (AC^{α}) is true but (DC^{ω}) is false and furthermore $On^{\mathfrak{M}} = On^{\mathfrak{N}}$ and $\mathfrak{M}$ and $\mathfrak{N}$ have the same alephs.

The Independence of the Axiom of Choice from the Principle of Dependent choices

Mostowski showed in 1948 that (AC) is independent from (DC) in ZF^0. W.Marek in Warsaw translated Mostowski's construction of a model $\mathfrak{M}$ of ZF^0 + (DC) + $\neg$ (AC^{ω_1}) to yield a Cohen-generic model $\mathfrak{N}$ of ZF plus (DC) + $\neg$ (AC). Thus $(DC^{\omega}) \to (AC)$ is not provable in full ZF.

[57] W.MAREK: A remark on independence proofs; Bull. Acad. Polon. Sci. vol.14(1966)p.543-545.

Marek just immitates Mostowski's model by adding to a given countable standard model $\mathfrak{M}$ of ZF + V = L generically a set X of $\aleph_1$ many unordered pairs B_{μ} $(\mu \in \omega_1)$ where each B_{μ} contains two Cohen-generic reals U_{μ}^0 and U_{μ}^1. The extension $\mathfrak{N}$ is obtained as the constructible closure (using ordinals of $\mathfrak{M}$), but besides Gödel's eight fundamental operations Marek takes a nineth one which will serve to add countable sequences. Unfortunately the proof in [57] is only briefly sketched.

We shall consider in the sequel Feferman's model $\mathfrak{M}[a_0, a_1, .. a_n ..]$. Feferman showed that in this model the (BPI) fails and Dana Scott pointed out that even the axiom of choice for unordered pairs (AC_2) is violated - see Feferman [16] (we cited this paper on p.98). Several people observed that (DC) holds in Feferman's model. Probably R.M.Solovay was the first who made this observation. A proof is published in:

[73] G.E.SACKS: Measure - Theoretic Uniformity in Recursion Theory and Set Theory; Transactions Amer.Math.Soc.vol.142(1969)p.381-420

The proof is presented in [73] in the language of measure-theoretic uniformity rather than in the forcing approach. The results of Sacks [73] are announced in: G.E.Sacks: Measure Theoretic Uniformity; Bull. Amer. Math. Soc. vol.73(1967)p.169-174, and (under the same title) in the Gödel-Festschrift (Springer-Verlag, Berlin 1969)p.51-57.

FEFERMAN'S MODEL $\mathfrak{M}[a_0, a_1, \ldots]$.

Let $\mathfrak{M}$ be a countable standard model of ZF + V = L. Define in $\mathfrak{M}$ a ramified language $\mathcal{L}$, which contains besides the usual ZF-symbols ($\neg, \vee, \bigvee$, =, ε and variables), limited existential quantifiers $\bigvee^\alpha$ and limited comprehension operators E^α (for ordinals α of $\mathfrak{M}$), constants $\underline{x}$ for each set x of $\mathfrak{M}$ and unary (generic) predicates $\dot{a}_i$ for $i \in \omega$. We suppose that this is done in such a way that the correspondences $x \to \underline{x}$, $i \to \dot{a}_i$, $\alpha \to \bigvee^\alpha$ and $\alpha \to E^\alpha$ are all $\mathfrak{M}$-definable. This can be attained e.g. by the standard-method - see page 79. Define conditions p to be finite partial functions from $\omega \times \omega$ into $2 = \{0,1\}$. The (strong) forcing relation $\Vdash$ between conditions p and $\mathcal{L}$-sentences is defined as usual. The definition contains the following key-clause:

$$p \Vdash \dot{a}_i(t) \Leftrightarrow (\exists n \in \omega)(p \Vdash t = \underline{n} \;\&\; p(\langle n,i\rangle) = 1)$$

where t is a constant term of $\mathcal{L}$. Obtain a complete sequence $\mathfrak{R}$ and thereby a valuation val(t) of the constant terms t of $\mathcal{L}$. Let $\mathfrak{N} = \{\text{val}(t);\ t \text{ a constant term of } \mathcal{L}\}$, then $\mathfrak{N}$ is a countable standard model of ZF. It holds that $a_i = \text{val}(\dot{a}_i) \subseteq \omega$. We use the following notation

$$\mathfrak{N} \triangleq \mathfrak{M}[a_0, a_1, \ldots]$$

i.e. $\mathfrak{N}$ results from $\mathfrak{M}$ by adding countably many Cohen-generic reals a_i ($i \in \omega$) to $\mathfrak{M}$. Notice, that the correspondence $a_i \to i$, id est $\{\langle a_i, i\rangle;\ i \in \omega\}$, is not added and that we did not add a set A which just collects these generic reals a_i. (Concerning the latter) This is the main difference to the model of Halpern-Lévy: $\mathfrak{M}[a_0, a_1, \ldots, A]$ (see pages 101-103 and p.131). Halpern-Lévy's model satisfies the (BPI) and since

$(BPI) \rightarrow (OP) \rightarrow (AC_2)$ (in ZF),

where (BPI) is the Boolean prime ideal theorem, (OP) the ordering principle and (AC_2) the statement which says that every set of unordered pairs has a choice-function, it follows, that (AC_2) holds in $\mathfrak{M}[a_0, a_1, \ldots, A]$. We shall show in the sequel that (AC_2) does not hold in $\mathfrak{M}[a_0, a_1, \ldots]$. Moreover it will be shown that there is no set in $\mathfrak{M}[a_0, a_1, \ldots]$ which just collects the reals $a_0, a_1, \ldots, a_n, \ldots$. Though the models of Feferman and of Halpern-Lévy seem to be very similar, they are considerably different and have extremely different features.

Symmetry-properties of Feferman's model

Let G be the group (in $\mathfrak{M}$) of all one-to-one mappings π of ω onto ω. Let $\pi(\Phi)$ be the result of substituting $\dot{a}_{\pi(i)}$ for $\dot{a}_i$ in the $\mathcal{L}$-formula Φ, and define $\pi(p) = \{\langle\langle n,\pi(i)\rangle,e\rangle;\ \langle\langle n,i\rangle,e\rangle \in p\}$ for conditions p. Then the "classical" Symmetry-lemma says:

"If $\pi \in G$, Φ is an $\mathcal{L}$-sentence and p a condition, then

$p \Vdash \Phi$ iff $\pi(p) \Vdash \pi(\Phi)$"

Feferman considers in [16] p.330-331, a different kind of transformation (we use Lévy's notation in [51] p.147-149).

If r is a set of $\mathfrak{M}$ and $r \subseteq \omega \times \omega$, then r defines a transformation:

Definition: Let Q be a function on a subset of $\omega \times \omega$ into 2 (in particular Q may be a condition). We define

$$[r,Q] = \{\langle\langle n,i\rangle,e\rangle ;\ (\langle\langle n,i\rangle,e\rangle \in Q \wedge \langle n,i\rangle \in r) \vee \vee(\langle\langle n,i\rangle,1-e\rangle \in Q \wedge \langle n,i\rangle \notin r)\}.$$

Definition. Let Φ be an $\mathcal{L}$-formula. We write $[r,\Phi]$ for the result of replacing each occurrence of $\dot{a}_i(\zeta)$ (where ζ is a variable or a constant term of $\mathcal{L}$) in Φ by

$$[r,\zeta] \in \omega \wedge (\dot{a}_i([r,\zeta]) \leftrightarrow \langle [r,\zeta], \underline{i}\rangle \in r),$$

where $[r,\zeta] \mathrel{\hat=} \zeta$ if ζ is a variable and $\langle [r,\zeta],\underline{i}\rangle \in r$ stands for:

$$\bigvee\nolimits_x^{\omega}[x \in r \wedge \bigwedge\nolimits_y^{\omega}(y \in x \leftrightarrow (\bigwedge\nolimits_z^{\omega}(z \in y \leftrightarrow z \simeq [r,\zeta]) \vee \bigwedge\nolimits_z^{\omega}(z \in y \leftrightarrow (z \simeq [r,\zeta] \vee z \simeq \underline{i}))))$$

(the Kuratowski-definition of an ordered pair as a limited sentence of $\mathcal{L}$) and $z \simeq [r,\zeta]$ for $\bigwedge\nolimits_x^{\omega}(v \in z \leftrightarrow v \in [r,\zeta])$.

It follows, that if p is a condition (in $\mathfrak{M}$) and $r \subseteq \omega \times \omega$, r in $\mathfrak{M}$, then $[r,p]$ is a condition (in $\mathfrak{M}$). Further if Φ is a limited $\mathcal{L}$-formula, then $[r,\Phi]$ is a limited $\mathcal{L}$-formula; if Φ is unlimited, then so is $[r,\Phi]$. If t is a limited comprehension term, say $E^{\alpha}x\Phi(x)$, then $[r,t] \triangleq E^{\alpha}x[r,\Phi(x)]$ is a limited comprehension term (this is used above for $[r,\zeta]$ if ζ is a constant term).

Lemma I. Let Q_1 and Q_2 be functions with $\mathrm{Dom}(Q_1) \subseteq \omega \times \omega$, $\mathrm{Dom}(Q_2) \subseteq \omega \times \omega$ and $\mathrm{Range}(Q_1) \subseteq 2$, $\mathrm{Range}(Q_2) \subseteq 2$ and let $r \subseteq \omega \times \omega$ and suppose that Q_1, Q_2 and r are in $\mathfrak{M}$. If $Q_1 \subseteq Q_2$, then $[r,Q_1] \subseteq [r,Q_2]$. Further it holds that $[r,[r,Q_i]] = Q_i$ (for i = 1,2). If $\Phi(x)$ is an $\mathcal{L}$-formula and $\Phi'(x) = [r,\Phi(x)]$, then $[r,\Phi(u)] = \Phi'([r,u])$ if u is a variable of $\mathcal{L}$ or a constant term of $\mathcal{L}$. Further it holds that $\delta[r,t] = \delta(t)$ for any constant term t of $\mathcal{L}$.

(see p.79 for definition of δ).

Lemma II. If $\mathcal{R} \triangleq (p_0, p_1, \ldots, p_n, \ldots)$ is a complete sequence of conditions and if $r \subseteq \omega \times \omega$ is a set in $\mathfrak{M}$, then

$$[r,\mathcal{R}] \triangleq_{\mathrm{Def}} ([r,p_0],[r,p_1],\ldots,[r,p_n],\ldots)$$

is a complete sequence.

Lemma III. For every constant term t: $\mathrm{val}_{\mathcal{R}}(t) = \mathrm{val}_{[r,\mathcal{R}]}([r,t])$ for r and $\mathcal{R}$ as in the preceeding lemma. For every $\mathcal{L}$-sentence ,

$$\mathfrak{N}_{\mathcal{R}} \models \Phi \quad \text{iff} \quad \mathfrak{N}_{[r,\mathcal{R}]} \models [r,\Phi].$$

For a detailed proof see Lévy's paper [51] p.148. Lemma III can be strengthened to:

Lemma IV. (Feferman [16]): Let Φ be an $\mathcal{L}$-sentence, p a condition (in $\mathfrak{M}$) and $r \subseteq \omega \times \omega$, r a set of $\mathfrak{M}$. Then

$$p \Vdash \Phi \quad \text{iff } [r,p] \Vdash [r,\Phi].$$

Next we shall present Feferman's lemma, which says, that (BPI) does not hold in $\mathfrak{N}$. We need some lemmata.

Lemma V. Let $P(\omega)$ be the powerset of ω and I a prime ideal in $P(\omega)$. If I is not principal, then I contains all finite subsets of ω.

Proof. Suppose, there is a finite subset $S = \{b_1, \ldots, b_n\} \subseteq \omega$ such that $S \notin I$. Then $\omega - S \in I$ since I is prime. Write $B_1 = \omega - \{b_1\}, B_2 = \omega - \{b_2\}, \ldots, B_n = \omega - \{b_n\}$. Then $\bigcap\{B_i; 1 \leq i \leq n\} \in I$ and since I is prime there exists i with $1 \leq i \leq n$ such that $B_i \in I$. Since I is a proper ideal, $I = \{x \subseteq \omega; x \subseteq B_i\}$. Hence I is a principal ideal,

q.e.d.

Lemma (Feferman [16] p.343): Let $P^{\mathfrak{N}}(\omega)$ be the powerset of ω in the sense of $\mathfrak{N}$. Every prime-ideal $I \in \mathfrak{N}$ in the Boolean algebra $\langle P^{\mathfrak{N}}(\omega), \subseteq \rangle$ is a principal ideal.

Proof. Suppose there exists in $\mathfrak{N}$ a non-principal prime ideal I of $\langle P^{\mathfrak{N}}(\omega), \subseteq \rangle$. Thus I contains all finite subsets of ω. Let t_I be a constant term of $\mathcal{L}$ such that $I \triangleq val(t_I)$ and $t_I \triangleq E^{\alpha}x\Phi(x)$.
Let $occ(\Phi) = \{i \in \omega; \dot{a}_i$ occurs in $\Phi\}$ and $n \in \omega$ such that $i \in occ(\Phi)$ implies $i < n$. We shall show that neither a_n nor $\omega - a_n$ is in I (a_n generic). Proceed by cases:

Case 1. Suppose $a_n \in I$ holds in $\mathfrak{N}$. Since everything that holds in $\mathfrak{N}$ is forced by some conditions in $\mathcal{R}$, the complete sequence which defines $\mathfrak{N}$, there exists $p \in \mathcal{R}$ such that

$$p \Vdash \Phi(\dot{a}_n).$$

Let k_0 be chosen so that for all $k \geq k_0$, $\langle k,n \rangle \notin Dom(p)$.
Define in $\mathfrak{M}$:

$$r = \{\langle k,m \rangle; m \neq n \vee (m = n \wedge k < k_0)\} \subseteq \omega \times \omega.$$

It follows that $[r,p] = p$. Let $\Phi'(x) = [r,\Phi(x)]$. Lemma IV implies $p \Vdash [r,\Phi(\dot{a}_n)]$, hence $\mathfrak{N} \models [r,\Phi(\dot{a}_n)]$ since $[r,p] = p \in \mathcal{R}$. But $\mathfrak{N} \models [r,\Phi(\dot{a}_n)] \leftrightarrow \Phi'([r,\dot{a}_n])$. By construction of r: $\mathfrak{N} \models \Phi'([r,\dot{a}_n]) \leftrightarrow \Phi([r,\dot{a}_n])$. Hence

$$\mathfrak{N} \models \Phi([r,\dot{a}_n])$$

and therefore $val([r,\dot{a}_n]) \in I$. But

$$val([r,\dot{a}_n]) = [(a_n \cap k_0) \cup (\omega - a_n)] \cap (\omega - k_0).$$

Since prime ideals J satisfy: $a \sqcap b \in J$ then $a \in J$ or $b \in J$, it follows that either $(a_n \cap k_0) \cup (\omega - a_n)$ or $(\omega - k_0)$ is in I. But I contains all finite subsets of ω, hence in particular k_0. Hence $\omega - k_0$ cannot be in I since otherwise $k_0 \cup (\omega - k_0) = \omega \in I$. Thus we get that $(a_n \cap k_0) \cup (\omega - a_n) \in I$. Hence $\omega - a_n \in I$. But by our assumption $a_n \in I$, a contradiction, ω would be in I.

Case 2. $\omega - a_n \in I$ holds in $\mathfrak{N}$. Proceed in a similar way and obtain a contradiction. This proves the lemma.

Corollary: The Boolean Prime Ideal Theorem does not hold in $\mathfrak{N}$.

Proof. It is well-known that (in ZF) the (BPI) is equivalent to the statement: "Every infinite Boolean algebra has a non-principal prime ideal" (see e.g. Tarski's abstract in the Bull. AMS 60(1954)p.390-391). It follows, hence, from the previous lemma, that (BPI) does not hold in $\mathfrak{N}$.

Remark: Dana Scott showed that in Feferman's model $\mathfrak{N}$ even there does not exist a choice set selecting reals from the cosets of the rationals in the reals - see Feferman's paper [16] p.343-344.

Let $C(\alpha,\dot{a}_{i_1},\dot{a}_{i_2},\ldots,\dot{a}_{i_k})$ be the $\mathfrak{M}$-set of all constant terms t of $\mathcal{L}$ such that $\delta(t) \leqslant \alpha$ and symbols $\dot{a}_j$ occur in t only for $j \in \{i_1,\ldots,i_k\}$. Let $\mathfrak{M}[\alpha,a_{i_1},\ldots,a_{i_k}]$ be the $\mathfrak{N}$-set of those sets (of $\mathfrak{N}$) which are denoted by members of $C(\alpha,\dot{a}_{i_1},\ldots,\dot{a}_{i_k})$. It is clear that $\mathcal{L}$ has a constant term $\underset{\sim}{t}(\alpha,\dot{a}_{i_1},\ldots,\dot{a}_{i_n})$ which denotes $\mathfrak{M}[\alpha,a_{i_1},\ldots,a_{i_k}]$. Define

$$C(\dot{a}_{i_1},\ldots,\dot{a}_{i_k}) = \bigcup\{C(\alpha,\dot{a}_{i_1},\ldots,\dot{a}_{i_k});\ \alpha \in On^{\mathfrak{M}}\}$$

and let $\underset{\sim}{t}(\dot{a}_{i_1},\ldots,\dot{a}_{i_k})$ be (an unlimited) constant term of $\mathcal{L}$ denoting

$$\bigcup\{\mathfrak{M}[\alpha,a_{i_1},\ldots,a_{i_k}];\ \alpha \in On^{\mathfrak{M}}\} = \mathfrak{M}[a_{i_1},\ldots,a_{i_k}]$$

Lemma. For each finite subset $\{i_1,\ldots,i_k\}$ of ω, $\mathfrak{M}[a_{i_1},\ldots,a_{i_k}]$ has an $\mathfrak{N}$-definable well-ordering.

For the proof use the techniques presented on pages 97-98, 138 and 158.

Lemma. Let $\Phi(x)$ be an $\mathcal{L}$-formula whose only free variable is x such that $occ(\Phi(x)) = \{i \in \omega;\ \dot{a}_i \text{ occurs in } \Phi(x)\} \subseteq k \in \omega$. Then for every condition p:

$$p \Vdash \bigvee\nolimits_x^{\alpha}\Phi(x) \leftrightarrow (\bigvee\nolimits_x^{\alpha}\Phi(x) \wedge x \varepsilon \underset{\sim}{t}(\alpha,\dot{a}_0,\dot{a}_1,\ldots,\dot{a}_k)).$$

The idea for the proof is the following: if u is a constant term of $\mathcal{L}$, say $E^{\beta}x\Psi(x)$, with $\beta < \alpha$, such that $\Phi(u)$ and u mentions (names for) generic reals $\dot{a}_j$ at most for $j \in \{0,1,\ldots,m\}$, then transform u into a term $u^* = E^{\beta}x\Psi^*(x)$ such that $\Phi(u^*)$ and u^* mention (names for) generic reals $\dot{a}_j$ at most for $j \in \{0,1,\ldots,k\} = k+1$. This can be achieved by replacing $a_k, a_{k+1},\ldots,a_m$ in u by pairwise disjoint subsets of a_k. More precisely one defines (in $\mathfrak{M}$) the following function r from constant terms to constant terms (assume $k \leqslant m$):

$$r(E^{\omega}x\dot{a}_i(x)) = E^{\omega}x\dot{a}_i(x) \quad \text{if } 0 \leqslant i < k,$$

$$r(E^{\omega}x\dot{a}_{i+k}(x)) = E^{\omega}x\dot{a}_k((m-k+1)\cdot x+\underline{i}) \quad \text{if } 0 \leqslant i \leqslant m-k,$$

$$r(E^{\omega}x\mathring{a}_{i+m+1}(x)) = E^{\omega}x\mathring{a}_{i+k+1}(x) \quad \text{if } 0 \leqslant i.$$

Extend r to act on all constant terms of $\mathcal{L}$ in the following way. If $E^{\gamma}x\Gamma(x)$ is any constant term of $\mathcal{L}$, then replace first in $\Gamma(x)$ every occurence of $\mathring{a}_j(x)$ by $x \in E^{\omega}z\mathring{a}_j(z)$ and call the resulting formula $\Gamma'(x)$. Then replace every occurrence of $E^{\omega}z\mathring{a}_j(z)$ in $\Gamma'(x)$ by $r(E^{\omega}z\mathring{a}_j(z))$ and call the resulting formula $r(\Gamma(x))$. Finally define $r(E^{\gamma}x\Gamma(x))$ to be $E^{\gamma}xr(\Gamma(x))$. With these definitions let u^* be $E^{\beta}xr(\Psi(x))$, id est $r(u)$. It follows from the construction, that u^* mentions (names of) generic reals $\mathring{a}_j$ at most for $j \in k+1$. Hence $val(u^*) \in \mathfrak{M}[a_0, a_1, \ldots, a_k]$. A symmetry argument shows that $\Phi(u^*)$ holds.

Lemma (R.Solovay): The axiom of dependent choices (DC^{ω}) holds in Feferman's model $\mathfrak{N} \triangleq \mathfrak{M}[a_0, a_1, \ldots, a_n, \ldots]$.

Outline of proof. Let $\Phi(x,y)$ be an $\mathcal{L}$-formula whose only free variables are x and y, such that if $\mathring{a}_j$ occurs in Φ, then $j < m$. Let us assume for simplicity that $m = 1$. Suppose $E^{\alpha}\langle x,y\rangle\Phi(x,y)$ defines in $\mathfrak{N}$ a binary relation R on a set s such that for all $x \in s$ there exists $y \in x$ with $\langle x,y\rangle \in R$ in $\mathfrak{N}$. We intend to find in $\mathfrak{N}$ a function f from ω into s such that for all $n \in \omega$, $\langle f(n),f(n+1)\rangle \in R$ and $f \in \mathfrak{M}(a_0, a_1)$.

By the previous lemma it holds in $\mathfrak{N}$ that the following two formulae are equivalent:

(1) $\bigwedge_x \bigvee_y [x \in \underset{\sim}{t}(\mathring{a}_0, \mathring{a}_1, \ldots, \mathring{a}_n) \rightarrow \Phi(x,y)]$,

(2) $\bigwedge_x \bigvee_y [x \in \underset{\sim}{t}(\mathring{a}_0, \mathring{a}_1, \ldots, \mathring{a}_n) \rightarrow (\Phi(x,y) \wedge y \in \underset{\sim}{t}(\mathring{a}_0, \mathring{a}_1, \ldots, \mathring{a}_{n+1}))]$.

Consider the following transformation r_{n+1}:

$$r_{n+1}(E^{\omega}x\mathring{a}_0(x)) = E^{\omega}x\mathring{a}_0(x),$$

$$r_{n+1}(E^{\omega}x\mathring{a}_{i+1}(x)) = E^{\omega}x\mathring{a}_1(2^{i+1}\cdot x + 2^{i} - 1) \quad \text{for } 0 \leqslant i \leqslant n,$$

$$r_{n+1}(E^{\omega}x\mathring{a}_{n+2}(x)) = E^{\omega}x\mathring{a}_1(2^{n+1}\cdot x + 2^{n+1} - 1),$$

$$r_{n+1}(E^{\omega}x\mathring{a}_{n+j}(x)) = E^{\omega}x\mathring{a}_{1+j}(x) \quad \text{for } 3 \leqslant j \in \omega.$$

This transformation r can be extended to act on the whole $\mathfrak{M}$-class of constant terms in the same way as it was done in the preceeding proof. A symmetry argument yields that

(3) $\bigwedge_x \bigvee_y [x \in r_{n+1}(\underset{\sim}{t}(\mathring{a}_0, \ldots, \mathring{a}_n)) \rightarrow \Phi(x,y)]$, and

(4) $\bigwedge_x \bigvee_y [x \in r_{n+1}(\underset{\sim}{t}(\mathring{a}_0, \ldots, \mathring{a}_n)) \rightarrow \Phi(x,y) \wedge y \in r_{n+1}(\underset{\sim}{t}(\mathring{a}_0, \ldots, \mathring{a}_{n+1}))]$

are equivalent (namely, apply r_{n+1} to (1) ↔ (2)). Notice that $r_{n+1}(\underset{\sim}{t}(\dot{a}_0,\ldots,\dot{a}_n))$ and $r_{n+1}(\underset{\sim}{t}(\dot{a}_0,\ldots,\dot{a}_{n+1}))$ mention (names of) generic reals $\dot{a}_j$ only for $j = 0$ or $j = 1$. Hence $\mathrm{val}(r_{n+1}(\underset{\sim}{t}(\dot{a}_0,\ldots,a_k)) \subseteq \mathfrak{M}[a_0,a_1]$ for $k \in \{n,n+1\}$. But $\mathfrak{M}[a_0,a_1]$ has an $\mathfrak{N}$-definable wellordering. Hence there is in $\mathfrak{N}$ a function $f : \omega \to s$ such that $f(0) \in \mathfrak{M}[a_0,a_1]$,

$f(n+1) \in \mathrm{val}(r_{n+1}(\underset{\sim}{t}(\dot{a}_0,\ldots,\dot{a}_{n+1}))) \subseteq \mathfrak{M}[a_0,a_1]$,

and $\bigwedge_{n\in\omega}\langle f(n),f(n+1)\rangle \in R$ holds in $\mathfrak{N}$. We can insist that $f \in \mathfrak{M}[a_0,a_1]$, since the constant-terms that denote the well-orderings needed in the definition of the term t_f (for $f = \mathrm{val}(t_f)$) are members of $C(\dot{a}_0,\dot{a}_1)$. This proves the lemma.

Theorem (R.Solovay): If ZF is consistent, then so is ZF + $\bigwedge_\alpha(AC^\alpha)$ + + (DC^ω) + $\neg\,(DC^{\omega_1})$.

The model used by Solovay is the Cohen-generic extension $\mathfrak{N} \triangleq \mathfrak{M}[a_0,\ldots,a_\gamma,\ldots]_{\gamma<\omega_1^{\mathfrak{M}}}$ of a countable standard model $\mathfrak{M}$ of ZF + V = L which results from $\mathfrak{M}$ by adding ω_1 many (in the sense of $\mathfrak{M}$) generic reals a_γ ($\gamma < \omega_1^{\mathfrak{M}}$) to $\mathfrak{M}$ but no set collecting these reals.

Another result in this area is due to Tomás Jech from Prague:

[38] T.JECH: Interdependence of weakened forms of the axiom of choice; Comment.Math.Univ.Carolinae, Prague, vol.7(1966)p.359-371, Corrections, vol.8(1967) page 567.

Theorem (J.Jech [38]): Let $\mathfrak{M}$ be a countable standard model of ZF + (AC) and α a regular infinite cardinal in $\mathfrak{M}$. Then there is an extension $\mathfrak{N}$ of $\mathfrak{M}$ with the same ordinals such that $\mathfrak{N}$ is a ZF-model satisfying (AC^β) and (DC^β) for every $\beta < \alpha$ but neither (AC^α) nor (DC^α) hold in $\mathfrak{N}$.

H.C.Doets asked, whether there is any interdependence between (DC^ω) and the (BPI). The answer follows from results presented in this chapter. In fact, Halpern and Lévy showed that $\mathfrak{M}[a_0,a_1,\ldots,A]$ satisfies (BPI) + $\neg\,(AC^\omega)$ and hence $\neg\,(DC^\omega)$ – see pages 100-103 and section G, p.131. On the other hand (DC^ω) + $\neg$ (BPI) hold in Feferman's model $\mathfrak{M}[a_0,a_1,\ldots]$.

I) A FINAL WORD

The reader will have observed that in numerous proofs of this chapter we referred mostly to the weak-forcing relation $\Vdash^*$ and in some few places only to strong forcing $\Vdash$. For this reason it is best to forget about strong forcing and introduce from the beginning only weak forcing (and then using $\Vdash$ for this weak forcing relation). If one wants to do so, then one defines in the given countable standard model $\mathfrak{M}$ a ramified language $\mathcal{L}^*$ having limited universal quantifiers $\bigwedge^\alpha$ rather than existential ones. As an example we consider the model $\mathfrak{M}[a_0, a_1, \ldots, A]$ of Halpern-Lévy. Then e.g. (if $\Vdash$ denotes weak forcing!)

$$(*) \quad p \Vdash t \varepsilon \dot{A} \Leftrightarrow (\forall p' \supseteq p)(\exists q \supseteq p')(\exists i \in \omega)(p \Vdash t \simeq \dot{a}_i).$$

Since complete sequences force weakly $\mathcal{L}^*$-sentences iff they do so strongly, the extensions obtained by weak or strong forcing do not differ. Though the weak-forcing approach is much smoother going, we presented strong forcing as the basic relation, since from the pädagogical point of view a definition like

$$(**) \quad p \Vdash t \varepsilon \dot{A} \Leftrightarrow (\exists i \in \omega)(p \Vdash t \simeq \dot{a}_i)$$

is much easier to understand than (*).

We intended to present in these lectures a selection of the most important techniques and results from the field of model theory of set theory. Measured by the number of pages of these notes and the time of one semester, I hope the reader will agree that the selection is not so small. But if we measure the selection by the number of results obtained so far we are (strongly) forced to say that we could present only some few results. It remains to refer to the litterature and to the following survey article:

[59] A.R.D.MATHIAS: A survey of recent results in Set Theory; mimeographed, Bonn-Stanford 1968. To appear in the UCLA-- Set Theory Symposium-Proceedings.

REFERENCES

[1] W.ACKERMANN: Zur Axiomatik der Mengenlehre; Math.Annalen 131,p.336 - 345 (1956).

[2] J.W.ADDISON: Some consequences of the axiom of constructibility; Fund.Math.46,p.337 - 357 (1959).

[3] J.L.BELL - A.B.SLOMSON: Models and Ultraproducts, an introduction; North-Holland Publ.Comp.Amsterdam-London 1969.

[4] M.BOFFA: Les ensembles extraordinaires; Bull.Soc.Math.Belgique 20,p.3 - 15 (1968).

[5] M.BOFFA - G.SABBAGH: Sur l'axiome Ü de Felgner; C.R.Acad.Sci.Paris Sér.A-B 270,p. A993 - A994 (1970).

[6] G.BOOLOS: On the semantics of the constructible levels; Zeitschr.f.math.Logik u.Grundl.d.Math.16,p.139-148(1970).

[7] G.BOOLOS - H.PUTNAM: Degrees of Unsolvability of constructible sets of integers; J.S.L. 33,p.497-513(1968).

[8] G.CANTOR: Gesammelte Abhandlungen (Edited by E.Zermelo), Springer-Verlag Berlin 1932 (Reprint:G.Olms,Hildesheim 1962).

[9] P.J.COHEN: The Independence of the axiom of choice; mimeographed notes (32 pages),Stanford University 1963.

[10] P.J.COHEN: The Independence of the Continuum Hypothesis; Proc. Nat.Acad.Sci.USA,Part I in vol.50,p.1143-1148(1963), Part II in vol.51,p.105-110(1964).

[11] P.J.COHEN: Independence Results in Set Theory; In: The Theory of Models - Symposium, North-Holland Publ.Comp.Amsterdam 1965, p.39 - 54.

[12] P.J.COHEN: Set Theory and the Continuum Hypothesis; Benjamin Inc.,New York - Amsterdam 1966.

[13] D.v.DALEN: Set Theory from Cantor to Cohen, to appear.

[14] W.B.EASTON: Powers of regular Cardinals; Annals of Math.Logic, 1, p.139 - 178 (1970).

[15] S.FEFERMAN: Arithmetization of metamathematics in a general setting; Fund.Math.49,p.35-92(1960).

[16] S.FEFERMAN: Some applications of the notions of forcing and generic sets; Fund.Math.56,p.325-345(1965).

[17] U.FELGNER: Die Inklusionsrelation zwischen Universa und ein abgeschwächtes Fundierungsaxiom; Archiv d.Math.20,p.561-566 (1969).

[18] U.FELGNER: Über das Ordnungstheorem; to appear in Zeitschr.f. math.Logik u.Grundl.d.Math.1971.

[19] U.FELGNER:Comparison of the axioms of local and universal choice; Fund.Math. 71 (1971). - see also J.S.L.35,p.603(1970).

[20] A.FRAENKEL: Der Begriff „ definit" und die Unabhängigkeit des Auswahlaxioms; Sitzungsberichte d.Preuss.Akad.Wiss.,Phys.Math. Klasse, 21, p.253-257(1922).

[21] A.FRAENKEL: Über eine abgeschwächte Fassung des Auswahlaxioms; J.S.L.2,p.1-27(1937) (Abstracts: C.R.Acad.Sci.Paris 192,p.1072 (1931), and: Jahresberichte d.DMV 41,p.88 of Part II (1931)).

[22] R.J.GAUNTT: Undefinability of Cardinality; Proceedings of the UCLA - Set Theory Institute 1967, to appear in: Proc.of Symp. in pure Math.,vol.13,Part II.

[23] K.GÖDEL: The Consistency of the Axiom of Choice and the Generalized Continuum Hypothesis; Proc.Nat.Acad.Sci.USA 24,p.556 - 557(1938).

[24] K.GÖDEL: Consistency-Proof for the Generalized Continuum Hypothesis;Proc.Nat.Acad.Sci.USA 25,p.220-225(1939).

[25] K.GÖDEL: The Consistency of the Axiom of Choice and of the Generalized Continuum-Hypothesis; Princeton Univ.Press 1966 (7th printing), Annals of Math.Studies Nr. 3 .

[26] K.GÖDEL: What is Cantor's Continuum Problem? Amer.Math.Monthly 54,p.515-525(1947), Corrections ibidem 55,p.151(1948).

[27] R.GREWE: On Ackermann's Set Theory; Dissertation, University of California Las Angeles 1966.

[28] R.GREWE: Natural models of Ackermann's Set Theory; J.S.L.34 p.481 - 488 (1969).

[29] P.HÁJEK: Syntactic models of Axiomatic Theories; Bull.Acad. Polon.Sci.13,p.273-278(1965).

[30] P.HÁJEK: Generalized Interpretability in terms of Models - Note to a paper of R.Montague; Casopis pro pestovani matematiki 91, p.352 - 357 (1966).

[31] J.D.HALPERN: The Boolean Prime Ideal Theorem; Lecture Notes UCLA-Set Theory Institute 1967(informally distributes manuscript)

[32] J.D.HALPERN: The independence of the axiom of choice from the Boolean Prime Ideal Theorem; Fund.Math.55,p.57-66(1964).

[33] J.D.HALPERN - H.LÄUCHLI: A Pertition Theorem; Transactions Amer.Math.Soc. 124, p.360 - 367 (1966).

[34] J.D.HALPERN - A.LÉVY: The Ordering-Theorem does not imply the axiom of choice; Notices AMS 11,p.56 (1964).

[35] J.D.HALPERN - A.LÉVY: The Boolean Prime Ideal Theorem does not imply the axiom of choice; In: Proceedings of Symp.in pure Math.vol.XIII,Part I,p.83-134 (AMS,Providence 1971).

[36] K.HAUSCHILD: Bemerkungen,das Fundierungsaxiom betreffend; Zeitschr.math.Logik u.Grundl.d.Math.12,p.51-56(1966).

[37] M.JAEGERMANN: The axiom of choice and two definitions of Continuity; Bull.Acad.Polon.Sci.13,p.699-704(1965).

[38] T.JECH: Interdependence of weakened forms of the axiom of choice; Comment.Math.Univ.Carolinae,Prague, 7,p.359-371(1966), Corrections ibidem 8, p.567(1967).

[39] R.B.JENSEN: Modelle der Mengenlehre; Springer Lecture Notes vol.37 (Berlin-Heidelberg 1967).

[40] R.B.JENSEN: Concrete Models of Set Theory; In: Sets,Models and Recursion Theory,Leicester Proceedings 1965, North-Holland Publ.Comp.Amsterdam 1967, p.44-74.

[41] B.JONSSON: Homogeneous universal relational systems; Math. Scand.8, p.137 - 142 (1960).

[42] C.KARP: A Proof of the relative Consistency of the Continuum Hypothesis; In: Sets,Models and Recursion Theory, Proceedings Leicester 1965,· North-Holland Publ.Comp.Amsterdam 1967.

[43] C.KARP: Languages with Expressions of Infinite Length; North-Holland Publ.Comp.Amsterdam 1964.

[44] W.KINNA -K.WAGNER: Über eine Abschwächung des Auswahlpostulates; Fund.Math. 42,p.75-82(1955).

[45] A.LÉVY: On Ackermann's Set Theory; J.S.L.24,p.154-166(1959).

[46] A.LÉVY: Axiom schemata of strong infinity in axiomatic set theory; Pacific J.Math. 10 ,p.223-238(1960).

[47] A.LÉVY: On a spectrum of set theories; Illinois J.Math. 4 , p.413-424(1960).

[48] A.LÉVY: A hierarchy of formulas in Set Theory; Memoirs of the Amer.Math.Soc. Nr.57 (Providence 1965).

[49] A.LÉVY: The Fraenkel-Mostowski Method for Independence Proofs in Set Theory; In: Symposium on the Theory of Models, (North-Holland Publ.Comp.Amsterdam 1965)p.221-228.

[50] A.LÉVY: The Definability of Cardinal Numbers; In: Foundations of Mathematics, Gödel-Festschrift, Springer-Verlag 1969, p.15-38.

[51] A.LÉVY: Definability in axiomatic Set Theory I; In: Logic, Methodology and Philosophy of Science, Congress Jerusalem 1964, North-Holland Publ.Comp.Amsterdam 1966, p.127-151.

[52] A.LÉVY: The Interdependence of certain consequences of the axiom of choice; Fund.Math.54, p.135-157(1964).

[53] A.LINDENBAUM - A.TARSKI: Communication sur les recherches de la théorie des ensembles; Comptes Rendus de Seances de la Soc. Sci.et lettres, Varsovie 19(Classe 3)p.299-330(1926).

[54] A.LINDENBAUM - A.MOSTOWSKI: Über die Unabhängigkeit des Auswahlaxioms und einiger seiner Folgerungen; C.R.Soc.Sci.Lettr. Varsovie, Classe 3, vol.31, p.27-32(1938).

[55] A.A.LJAPUNOW - E.A.STSCHEGOLKOW - W.J.ARSENIN: Arbeiten zur Deskriptiven Mengenlehre; VEB Deutscher Verlag der Wiss., Berlin 1955.

[56] A.LÉVY - R.L.VAUGHT: Principles of partial reflection in the set theories of Zermelo and Ackermann; Pacific J.Math.11, p.1045 - 1062 (1961).

[57] W.MAREK: A remark on independence proofs; Bull.Acad.Polon.Sci. 14, p.543-545(1966).

[58] A.R.D.MATHIAS: The Order-Extension Principle; Proceedings of the 1967-set theory symposium at UCLA, vol.2, to appear.

[59] A.R.D.MATHIAS: A survey of recent results in Set Theory; Proc. of the 1967-set theory symposium at UCLA, to appear.

[60] E.MENDELSON: Introduction to Mathematical Logic; v.Nostrand Company, Princeton-New York-London 1966 (3rd printing).

[61] E.MENDELSON: Some Proofs of Independence in Axiomatic Set Theory; J.S.L. 21, p.291-303 (1956).

[62] E.MENDELSON: The Independence of a weak axiom of choice; J.S.L. 21, p.350-366(1956).

[63] R.MONTAGUE: Fraenkel's Addition to the axioms of Zermelo; Fraenkel-Festschrift: „Essays on the Foundations of Mathematics" The Magnus Press, Jerusalem 1967 (2nd edition).

[64] A.MOSTOWSKI: Über die Unabhängigkeit des Wohlordnungssatzes vom Ordnungsprinzip; Fund.Math.32,p.201-252(1939).

[65] A.MOSTOWSKI: On models of ZF-Set Theory satisfying the axiom of constructibility; Acta Philosophica Fennica, Fasc.18 , p.135 - 144 (1965).

[66] A.MOSTOWSKI: Constructible Sets, with applications; North-Holland Publ.Comp. + PWN, Amsterdam - Warszawa 1969.

[67] A.MOSTOWSKI: On a Problem of W.Kinna and K.Wagner; Colloquium Math. 6, p.207 - 208 (1958).

[68] A.MOSTOWSKI: On the principle of dependent choices; Fund.Math. 35, p.127 - 130 (1948).

[69] H.PUTNAM: A note on constructible sets of integers; Notre Dame J.Formal Logic 4,p.270-273(1963).

[70] H.RASIOWA - R.SIKORSKI: The Mathematics of Metamathematics, Monografie Matematyczne 41, Warszawa 1963.

[71] W.N.REINHARDT: Ackermann's Set Theory equals ZF; Annals of math.Logic 2, p.189 - 249 (1970).

[72] J.B.ROSSER: Simplified Independence Proofs; Academic Press New York - London 1969.

[73] G.E.SACKS: Measure-Theoretic Uniformity in Recursion Theory and Set Theory; Transactions AMS 142,p.381-420(1969).

[74] D.SCOTT: The notion of rank in set-theory; Summaries Summer Institute for Symbolic Logic,Cornell Univ.1957,p.267-269.

[75] J.R.SHOENFIELD: Mathematical Logic; Addison-Wesley Publ.Comp. Reading,Mass.-London 1967.

[76] J.R.SHOENFIELD: The problem of predicativity; Fraenkel-Festschrift,, Essays on the Foundations of Mathematics" The Magnus Press Jerusalem 1967 (2nd printing).

[77] J.R.SHOENFIELD: Unramified forcing; Proceedings of Symposia in Pure Math.vol.13,part I,p.357-381(AMS,Providence 1971).

[78] W.SIERPINSKI: Les ensembles projectifs et analytiques; Memorial des Sci.Math.,fasc.112, Paris (Gauthier-Villars) 1950.

[79] W.SIERPINSKI: Hypothèse du Continu. Warszawa-Lwow 1934 , 2nd edition: New York 1956.

[80] J.SILVER: Forcing à la Solovay; unpublished lecture notes.

[81] R.M.SOLOVAY: $2^{\aleph_0}$ can be anything it ought to be; In: The Theory of Models, 1963 Symposium at Berkely, North-Holland Publ.Comp.Amsterdam 1965, p.435.

[82] E.P.SPECKER: Zur Axiomatik der Mengenlehre (Fundierungs- und Auswahlaxiom); Zeitschr.f.math.Logik u.Grundl.d.Math. 3 , p.173 - 210 (1957).

[83] G.TAKEUTI: Topological Space and Forcing; (Abstract), J.S.L. 32,p.568-569(1967).

[84] E.J.THIELE: Über endlich axiomatisierbare Teilsysteme der ZF - Mengenlehre; Zeitschr.f.math.Logik u.Gr.d.Math. 14 , p.39 - 58 (1968).

[85] A.TARSKI - A.MOSTOWSKI - R.M.ROBINSON: Undecidable Theories; North-Holland Publ.Comp.Amsterdam 1953.

[86] P.VOPĚNKA: General Theory of ∇-Models; Comment.Math.Univ. Carolinae,Prague, 8, p.145 - 170 (1967).

[87] Hao WANG: Arithmetic translations of axiom systems; Transactions Amer.Math.Soc.71,p.283-293(1951).

[88] E.ZERMELO: Untersuchungen über die Grundlagen der Mengenlehre; Math.Annalen 65,p.261 - 281 (1908).

[89] E.ZERMELO: Über den Begriff der Definitheit in der Axiomatik; Fund.Math.14,p.339 - 344 (1929).